Video Cataloguing

Structure Parsing and Content Extraction

Video Cataloguing

Structure Parsing and Content Extraction

Guangyu Gao
Chi Harold Liu

CRC Press
Taylor & Francis Group
Boca Raton   London   New York

CRC Press is an imprint of the
Taylor & Francis Group, an **informa** business

CRC Press
Taylor & Francis Group
6000 Broken Sound Parkway NW, Suite 300
Boca Raton, FL 33487-2742

© 2016 by Taylor & Francis Group, LLC
CRC Press is an imprint of Taylor & Francis Group, an Informa business

No claim to original U.S. Government works

Printed on acid-free paper
Version Date: 20150904

International Standard Book Number-13: 978-1-4822-3577-7 (Hardback)

Visit the Taylor & Francis Web site at
http://www.taylorandfrancis.com

and the CRC Press Web site at
http://www.crcpress.com

Contents

Preface ix

Acknowledgment xiii

Authors xv

1 Introduction 1
1.1 Introduction to Movie and Teleplay Cataloguing 1
1.2 Related Research State and Progress 3
    1.2.1 Related Work on Shot Boundary Detection 6
    1.2.2 Related Work on Scene Detection and Recognition 7
    1.2.3 Related Work on Video Text Recognition 9
    1.2.4 Related Work on Character Identification 9
1.3 Main Research Work 11

2 Visual Features Extraction 13
2.1 Introduction 13
2.2 Scale-Invariant Feature Transform 14
2.3 Gabor Feature 17
2.4 Histogram of Oriented Gradients (HOG) 19
2.5 Maximally Stable Extremal Regions 20
2.6 Local Binary Pattern (LBP) 21
2.7 Feature Learning 22
2.8 Summary 23

3 Accelerating Shot Boundary Detection 25
3.1 Introduction 25
3.2 Related Work 27
3.3 Frame Difference Calculation 29
    3.3.1 Focus Region 29
    3.3.2 Frame Difference Measurement 30
3.4 Temporal Redundant Frame Reduction 33
3.5 Corner Distribution-Based MCFB Removal 34

3.6 Experimental Results...37
3.7 Summary...39

4 Key Frame Extraction 41
4.1 Introduction..41
4.2 Size of the Key Frame Set...42
4.3 Categories of Key Frame Extraction Methods.................43
4.4 Key Frame Extraction Using a Panoramic Frame 46
4.5 Summary...48

5 Multimodality Movie Scene Detection.................................49
5.1 Introduction..49
5.2 Related Work..50
5.3 KCCA and Feature Fusion-Based Method 52
    5.3.1 Shot Boundary Detection 52
    5.3.2 Key Frame Extraction 52
    5.3.3 Audiovisual Feature Extraction 53
    5.3.4 KCCA-Based Feature Fusion...............................54
    5.3.5 Scene Detection Based on the Graph Cut............56
5.4 Experiment and Results..58
5.5 Summary...59

6 Video Text Detection and Recognition 61
6.1 Introduction..61
6.2 Implementation of Video Text Recognition 62
    6.2.1 Video Text Detection 62
    6.2.2 Video Text Extraction 70
6.3 Summary...73

7 "Where" Entity: Video Scene Recognition............................75
7.1 Introduction..75
7.2 Related Work..77
7.3 Overview...78
7.4 Video Segmentation 79
7.5 Representative Feature Patch Extraction...........................79
7.6 Scene Classification Using Latent Dirichlet Analysis 81
7.7 Enhanced Recognition Based on VSC Correlation 83
7.8 Experimental Results...85
    7.8.1 Comparison of Different Methods........................85
    7.8.2 Performance Evaluation on Using Panoramic Frames..........86
    7.8.3 Comparison of Different Experimental Conditions 87
7.9 Summary...89

8 **"Who" Entity: Character Identification** **91**

    8.1   Introduction...91

    8.2   Introduction...93

    8.3   Related Work...95

    8.4   Overview of Adaptive Learning 96

    8.5   Adaptive Learning with Related Samples 100

        8.5.1   Review of LapSVM...100

        8.5.2   Related LapSVM 101

        8.5.3   Classification Error Bound of Related LapSVM...............104

    8.6   Experiments 106

        8.6.1   Database Construction 106

            8.6.1.1   Image Data 106

            8.6.1.2   Video Data 106

            8.6.1.3   Feature for Face Recognition 110

        8.6.2   Experiment Settings..110

        8.6.3   Video Constraint in Graph..111

        8.6.4   Related Sample in Supervised Learning 111

        8.6.5   Related Samples in Semisupervised Learning..................113

        8.6.6   Learning Curves of Adaptive Learning...........................115

        8.6.7   YouTube Celebrity Dataset 116

    8.7   Summary..118

9 **Audiovisual Information-Based Highlight Extraction** **119**

    9.1   Introduction...119

    9.2   Framework Overview 120

    9.3   Unrelated Scene Removal...121

        9.3.1   Three Types of Scene 121

        9.3.2   Speech and Nonspeech Discrimination 122

        9.3.3   Background Differences in SS and GS...............................123

    9.4   Unrelated Scene Removal...124

        9.4.1   Subshot Segmenting 125

        9.4.2   Score Bar Change Detection...125

        9.4.3   Special Audio Words Detection..126

    9.5   Experimental Results...126

    9.6   Summary..127

10 **Demo System of Automatic Movie or Teleplay Cataloguing**...............**129**

    10.1   Introduction...129

        10.1.1   Application Scenario..130

        10.1.2   Main Functions of the Demo 131

10.2 General Design of the Demo..132
    10.2.1 Overall Diagram of the Demo 132
    10.2.2 Running Environment..133
    10.2.3 Main Function Modules..133
    10.2.4 System Design and Realization 136

References...**139**
Index..**153**

Preface

Digital asset management (DAM) consists of management tasks and decisions surrounding the ingestion, annotation, cataloguing, storage, retrieval, and distribution of digital assets. Meanwhile, media asset management (MAM) is a subcategory of DAM that refers to digital assets of photographs, animations, videos, and audio. Media assets, in a broad sense, refer to all assets related to the media enterprise, such as office equipment, funds, and also other soft resources. But, in a narrow sense, media assets are just digital assets, namely, valuable digital information and data that are accumulated by media enterprises in their business and production processes.

The so-called media asset management is a comprehensive management solution on various types of media and content (such as video and audio data, text files, and pictures). Its core purpose is to make the media assets permanent through recycling and commercialization. So, what is the goal of media asset management? IBM Corporation has provided a simple construction goal for MAM; the key points include maximizing the value of the assets; reducing the cost of material classification, retrieval, and storage; providing cross-enterprise acquisition and reusability; offering better safety service; flexibly adapting to technical and business developments and changes; and making media asset management centralized and unified. In fact, we attempt to narrow the sense of MAM in the broadcast and television domain in this book. Specific to the broadcast and television domain, it mainly refers to the amount of movie and teleplay videos, as well as some sports videos.

In fact, the increase of availability of digital videos has led to the need for better methods of utilizing the content of video data in MAM. Unlike other information media, such as books (which have textual information), video data contain much richer information. Actually, with the rapidly accumulating number of TV programs and the rapid growth of online digital videos, it is crucial to be able to quickly retrieve the required data in MAM within massive videos. Actually, while *cataloguing* appeared in the definition of DAM, it is also the key issue for MAM. That is, with valid and effective cataloguing, the MAM system can effectively manage the annotation, storage, retrieval, and distribution of media. Therefore, video cataloguing is one of the most efficient ways to realize efficient and accurate media management in MAM.

Specifically, video cataloguing is a process to sort and refine videos in MAM, that is, to extract valuable images, fragments, and semantic words as sole files for video retrieval and recycling. Video cataloguing techniques have been proposed for years to offer people comprehensive understanding, convenient browsing, and retrieval of the whole video story. Generally, video cataloguing first extracts and refines a video's structure and semantic information and then configures the retrieval issues for video reutilization.

Besides, the advent of the digital age has created an urgent need to be able to store, manage, and digitally use historical and new video and audio material. Thus, cataloguing has emerged as a requirement of the times. Cataloguing holds an important position in media asset management, which is mainly to collect and refine the valuable fragments or elements in video and audio materials. Video cataloguing makes it easy to retrieve, reuse, and interact with the video resources, and then realize the maintenance and increment of the value of media assets. The catalogue is compiled based on data or information description. Namely, cataloguing is to analyze, select, and recode the media asset's form and content features first, and then to organize them in an orderly fashion according to certain rules for the data retrieval and query demands of users. Video cataloguing information, which can directly reflect the video structure and content, is the structural representation of the unstructured data. Actually, video cataloguing has always occupied a very important position in video content analysis, and it is also an indispensable way for the content provider to offer effective management. Generally, video structure analysis and semantic extraction are the foundations for the realization of automatic video cataloguing, and automatic video cataloguing is also the purpose or goal of video structure analysis and semantic extraction. Therefore, video cataloguing and video structure analysis and semantic extraction are mutually dependent. For example, through video structure analysis and semantic extraction, the extracted video structures and basic semantic data of movie and teleplay videos can be used as the cataloguing entry or item for automatic video cataloguing. At the same time, the goal of cataloguing is to effectively organize and manage video data. Hence, content-based video cataloging is also the premise and basis of content-based video retrieval, while video retrieval is the basic function of a video database or media asset.

A movie or teleplay is one of the indispensable consumption video resources in people's daily lives. Video resources don't have a specific structure or shape, but they do have a comprehensive logical structure in semantic content. Meanwhile, movie or teleplay videos are time-based media, except with common image characteristics, and they also have time and movement characteristics. Generally speaking, movie and teleplay videos can be described or interpreted from the following aspects:

1. **Video metadata.** Video metadata mainly refers to the global information (title, subject, type, etc.) and video filming and production information (producer, director, actor table, etc.).
2. **Video structure data.** Video structure data refers to the logical structure between video frames and video fragments. According to the broadcasting

standards, a whole video's structure can be divided into four levels from top to bottom: video program, video scene, video shots, and frame.

3. **Video semantic data**. Semantic data means the video semantic content descriptions, including the basic description of the video content extracted from video frames, audio, subtitles, and so forth (including captions, characters, time, place, action, and scenes).

4. **Video story**. If video semantic data are looked at as a series of discrete semantic words, then the video story is the organic combination of these words to form a comprehensive story, according to the movie's or teleplay's arranged grammar and content. Meanwhile, a video story is some type of natural language description. It accords with human thinking and understanding of habit to express the video's content using its story mode.

In fact, video metadata are the simple and global description of a whole movie or teleplay, which is determined in advance in the video script writing or before the video filming and production. Meanwhile, although the video story description can better meet people's watching or consumption demand, it is directly derived from the angle of human natural language. In reality, it is still difficult or challenging with current technologies for the machine to capture exactly the same with human language description. Therefore, this book mainly focuses on the research and analysis of the second- and third-level descriptions, namely, video structure data and video semantic data. By in-depth analysis of the video structural features, we can efficiently parse the video into meaningful units. After that, by acquiring a large number of basic semantic words, we can realize video summarization in the broadcast domain well. Finally, with this structural information and basic semantic words, movies and teleplays can be efficiently indexed, summarized, catalogued, and retrieved. It also brings forth a more efficient and reasonable method for video storage, management, and transmission.

In general, a movie or teleplay video is mapping into four levels of structure, from large to small: video, scene, shot, and frame. A shot is a basic physical unit of a video, while a scene is the basic logical or semantic unit. Thus, a shot is a physically existing unit, whereas a video scene logically exists. The use of video structure information can realize video storage, management, and transmission more efficiently, but from the user's point of view, it does not provide much help for a watching consumption experience. Therefore, it is necessary to get the basic semantic words for understanding or consuming these movies and teleplays. In addition, a story generally includes four elements: character, time, place, and event, of which characters are the most critical element, and the place is the context information of the whole story. Therefore, our research on basic semantic extraction focuses on extracting names of characters and categories of locations. In addition, while there are many captions in video frames, the video text extracted from captions should be suitable semantic data for video understanding and cataloguing.

Organization of the Book

We focus on dealing with the issues of video cataloguing, including video structure parsing and basic semantic word extraction, especially for movie and teleplay videos. The main goal is to obtain the structure and basic semantic data, in the hope of using these data to better serve video automatic cataloguing, indexing, and retrieval. Our starting point is to provide a fundamental understanding of video structure parsing; that is, we first discuss the video shot boundary detection, and then do research on video scene detection. In addition, we also address some basic ideas for semantic word extraction, including video text recognition, scene recognition, and character identification. Finally, a movie and teleplay cataloguing prototype is proposed.

This book is divided into 10 chapters. Chapter 1 introduces the main research contents and work of this book.

Chapter 2 discusses extraction as the primary processing. We will list and introduce some of the popularly used features in video analysis, especially visual features, the research background, and the present situation.

In Chapter 3, the most popular shot boundary detection methods are introduced and analyzed, comprising the accuracy and efficiency of shot boundary detection. Then, a novel accelerated shot boundary detection approach is proposed and compared with the existed methods to verify its efficiency and accuracy.

Chapter 4 covers key frame extraction, a very crucial step in video processing and indexing. In it we depict methods for key frame extraction and also propose a novel key frame extraction approach: the panoramic frame-based key frame extraction.

In Chapter 5, in order to feasibly browse and index movie and teleplay videos, we discuss and present research on movie scene detection as another important and critical step for video cataloguing, video indexing, and retrieval.

Since video text often contains plentiful semantic information, in Chapter 6 we propose how to efficiently extract video text as semantic data for cataloguing.

In Chapter 7, we propose a robust movie scene recognition approach based on a panoramic frame and representative feature patch. When our proposed approach was implemented to recognize scenes in realistic movies, the experimental results showed that it achieved satisfactory performance.

Chapter 8 describes how to recognize characters in movies and TV series accurately and efficiently, and then use these character names as cataloguing items for an intelligent catalogue.

Chapter 9 proposes the interesting application of highlight extraction in basketball videos. This application takes into consideration the importance of video text extraction.

Finally, in Chapter 10, we design and implement a prototype system of automatic movie and teleplay cataloguing (AMTC) based on the approaches introduced in the previous chapters.

Acknowledgment

We would like to thank the National Natural Science Foundation of China (grant no: 61401023) and the Fundamental University Research Fund of BIT (grants (2_2050205)_20130842005 and (2_2050205)_20140842001).

Authors

Guangyu Gao PhD, is an assistant professor at the School of Software, Beijing Institute of Technology, Beijing, China. He earned his PhD degree in computer science and technology from the Beijing University of Posts and Telecommunications, Beijing, China, in 2013, and his MS degree in computer science and technology from Zhengzhou University, Henan Province, China, in 2007. He was also a government-sponsored joint PhD student at the National University of Singapore, from July 2012 to April 2013. His current research interests include applications of multimedia, computer vision, video analysis, machine learning, and Big Data. As an author,

he received the Electronic Information and Science Technology Prize awarded by the Chinese Institute of Electronics. He has published dozens of prestigious conference and journal papers. He is a member of both the Association for Computing Machinery (ACM) and China Computer Federation.

Chi Harold Liu is a full professor at the School of Software, Beijing Institute of Technology, Beijing, China. He is also the director of the IBM Mainframe Excellence Center (Beijing), the IBM Big Data Technology Center, and the National Laboratory of Data Intelligence for China Light Industry. He earned a PhD degree from Imperial College, London, UK, and a BEng degree from Tsinghua University, Beijing, China. Before moving to academia, he joined IBM Research–China as a staff researcher and project manager, after working as a postdoctoral researcher at Deutsche Telekom Laboratories, Berlin, Germany, and a visiting scholar at IBM T. J. Watson Research Center, Hawthorne, New York. His current research interests include the Internet-of-Things (IoT), Big Data analytics, mobile computing, and wireless ad hoc, sensor, and mesh networks. He received the Distinguished Young Scholar Award in 2013, IBM First Plateau Invention Achievement Award in 2012, and IBM First Patent Application Award in 2011 and was interviewed by EEWeb.com as the featured engineer in 2011. He has published more than 60 prestigious conference and journal papers and owns more than 10 EU/U.S./China patents. He serves as the editor for *KSII's (Korean Society for Internet Information) Transactions on Internet and Information Systems* and the book editor of four books published by Taylor & Francis Group, Boca Raton, Florida. He also has served as the general chair of the SECON (Sensor and Ad Hoc Communications and Networks) '13 workshop on IoT networking and control, the IEEE WCNC (Wireless Communications and Networking Conference) '12 workshop on IoT enabling technologies, and the ACM UbiComp (Ubiquitous Computing) '11 workshop on networking and object memories for IoT. He has served as a consultant to the Asian Development Bank, Bain & Company, and KPMG, San Francisco, California, and as a peer reviewer for the Qatar National Research Foundation and National Science Foundation, China. He is a member of IEEE (Institute of Electrical and Electronics Engineers) and ACM.

Chapter 1

Introduction

In this chapter, we first describe the research background and significance of video cataloguing. While the definition of cataloguing in this book is mainly derived from broadcast program production and video resources management in the broadcast television domain, we introduce the related concepts of video structure parsing and basic semantic content extraction, as well as automatic cataloguing. Next, we review the research status and development of the structure analysis and basic semantic content extraction. Then we introduce the main content of this book specifically and separately. Finally, the organization of the book is presented.

1.1 Introduction to Movie and Teleplay Cataloguing

With the rapid development of multimedia technology and computer processing ability, people are facing a huge digital "information ocean." At the same time, broadcast and network video resources have already become a very important part of people's daily lives. Therefore, how to quickly and accurately classify and retrieve has become an urgent problem. In the field of TV broadcasting, most data are stored on magnetic tape in both analogue and digital form and this has resulted in several challenges and problems, such as the data being difficult to store, retrieve, and share. Recently, in order to efficiently index, store, and retrieve these data, most of the multimedia content is parsed and catalogued manually for structural and semantic cataloguing. Like the spurt of the number of videos (especially movies and teleplays), this traditional manual parsing and catalogue pattern, which needs a lot of manpower and material resources and has very low efficiency, is finding it harder and harder to meet the demand of movie and television program production and management. In view of the present situation of broadcast television for resource management and cataloguing, it has become urgent to provide a new mode to solve these problems. Actually, the media asset management system was generated in response to these problems and needs. In a media asset management system, video structure analysis, automatic and intelligent understanding, and cataloguing are the most important components. Automatic video parsing and cataloguing will undoubtedly save a lot of resources and improve the program efficiency in media asset management and each section in broadcast television production.

In fact, for automatic video structure analysis, understanding, and cataloguing, the crucial and first step is to understand the structure of movie and TV programs. Based on "Broadcast and Television Information Cataloguing Specification: Part 1: Specification for the Structure of Teleplays" [184], movies and TV programs should be divided into four levels in structure organization: the program layer, story level, scenario/scene layer, and shot layer. Thus, in order to realize the automatic video structure parsing, the corresponding technologies are needed, such as shot boundary detection and scene detection. Shot is defined as a continuous sequence of frames captured using one camera [26], and adjacent shots are connected through different conversion modes. Every video is composed of a plurality of shots, and the processing to detect the shots' connection position is called video shot boundary detection [150]. In addition, a video scene refers to a series of continuous video and audio materials with invariant visual background or scene appearances. Generally, a scene is composed of a plurality of shots with spatial and temporal consistency. Therefore, in a broad sense, a scene is a group of shots with the same semantic theme. Combining a series of shots with compressed content similarity together is called scene detection or scene segmentation, also known as shot clustering or logical story unit segmentation.

Using shot boundary detection and scene detection technology, one can effectively realize the automatic video structure parsing well. However, ensuring these related video structure parsing methods accurately and reasonably find all shot boundaries and scene boundaries is still a challenging problem. At the same time, according to specific video resources of movies and teleplays, multimodal information should be effectively used for more efficient and reasonable analysis of automatic video structure.

In addition, after automatic video structure parsing, where the movie or teleplay is split into several independent structural units (shots) and semantic units (video scenes), a consequential issue is how to understand these fragments and access high-level semantic content for semantic annotation and automatic cataloguing. Actually, this is one of the most crucial issues for the realization of video content transition from low-level features to high-level semantics across the semantic gap. The intuitive idea is to extract video text from captions in frames. Because the caption text is somehow the compact representation of the frame or video clip, video text recognition is implemented to offer the most direct semantic words for video cataloguing.

Except for direct video text, with the definition in broadcast television cataloguing rules, a video scene refers to a series of continuous video and audio materials with invariant visual background or scene appearances. Thus, understanding the unified and fixed background can also provide the basic place or location semantic content for automatic video cataloguing. Furthermore, a character is the basic substance in movies and teleplays as a key element to the story plot. Namely, a movie or teleplay unfolds around television characters and their activities most of the time. Therefore, character recognition and name labeling is a reasonable process to be used for automatic cataloguing in line with the audience habits. In recent years, face recognition technology for character recognition has gained hitherto unknown progress and has been widely used in research, industry, and real life. Thus, intuitively, the use of existing face recognition techniques can achieve good performance for automatic character identification in movies and teleplays. However, traditional face recognition technologies are usually required for recognition, with the face having obvious good appearance features. In movies and teleplays, a character's appearance always has a wide variety of changes and is also affected by the shooting environment, video quality, large-scale video amount, and other factors. As a result, traditional face recognition methods cannot be directly applied to character identification in movies and teleplays, and we need new approaches to realize the automatic recognition and label names of characters in them.

To sum up, for intelligent and automatic video cataloguing in the broadcast television domain, serious studies on automatic video structure parsing, video text, characters, locations, and other basic semantic extraction will greatly improve the efficiency of program production, transmission, broadcasting, and management. These studies also have a very important theoretical value and broad application prospect for multimedia management. Hence, automatic video cataloguing, including video structure parsing and basic semantic data extraction, has become a very important research topic, that is, the key content of video resource storage, transmission, and management. For example, movie and teleplay videos usually contain a huge number of audio and video materials. If these materials can be appropriately segmented and indexed, new valuable applications and video clip sources will be generated with retrieval and reuse through these video fragments and semantic information. A typical application involves the production of content, network transmission, broadcast, video indexing, browsing, and abstraction. For video content owners, including television stations, news media producers, and content producers, because traditional resource management is often carried out on the whole video with low efficiency and convenience, video structure parsing and automatic cataloguing technologies will enable more reasonable, effective, and efficient multimedia resource management.

1.2 Related Research State and Progress

Since the early 1970s, people have carried out research on image databases, and the realization formed was to support image queries with metadatabases constructed with manual input image attributes. However, with the development of multimedia technology in the 1990s, the amount of available image and other multimedia data has become much larger, and the volume of the database is increasing unceasingly. Therefore, the traditional method exposed its shortcomings. One of the those is that the manual input and annotation pattern consumed a large amount of man-power, especially for large-scale multimedia information databases, such as video databases, web network source databases, and digital libraries. In these information environments, there are plenty of new materials added every day, and all these newly added materials need to be archived in a timely manner. Without automatic or computer-aided processing, it is intolerant and source-consuming and cannot satisfy the cyclically update needs of users. Another shortcoming is that the audio and video data include abundant and comprehensive contents, and thus manual annotation has difficulty describing these rich contents clearly. The third shortcoming is that for real-time broadcast streaming processing, manual processing is totally infeasible. Thus, video cataloguing will solve these shortcomings by offering automatic or semiautomatic assistances.

For movie and teleplay videos, video cataloguing, including video structure analysis and basic semantic content extraction, aims to efficiently describe, store, organize, and index those videos of user interest [11,44,160]. In fact, video cataloguing involves many related technologies, such as databases, data mining, information retrieval, and video processing and analysis. Besides, video cataloguing is also related to content-based video retrieval (CBVR). In other words, it is the basic work or foundation for semantic video analysis and retrieval. CBVR analyzes and understands physical and semantic contents in multimedia data (such as video and audio streams) through computer assistance. Its essence is to make disordered video data structured and also to extract the semantic information to ensure the multimedia content can be quickly retrieved [48,172,219,229].

Since the 1990s, many scholars have conducted in-depth research on video content analysis and information retrieval, such as IBM's Query by Image Content (QBIC) project [59], ADVENT at Columbia University [170], and the Informedia project [138,183] at Carnegie Mellon University. After more than 20 years, CBVR has made considerable progress in many aspects, including video feature extraction and expression, video structurization, video cataloguing, and video indexing. Meanwhile, various retrieval and summarization systems have been proposed, such as MediaMill, QBIC, Informedia, CueVideo, and ADVENT. These systems provide video querying and indexing according to low-level features, sample images or clips, and keywords. But the current machines' intelligence is far less than that of the human brain; it is still relatively weak for machine understanding and recognition of complex contents. That is, the machine intelligence still cannot map video low-level features to high-level semantics [219]—the semantic gap problem. Actually, this problem has also hindered further development of CBVR. Therefore, the key issue regarding whether video retrieval can become an ideal application in daily life is if we can effectively establish an association between the level features and high-level semantics and also extract accurate video semantic information.

Among a large number of studies on video processing and analysis, the most concentrated area is analysis and processing of news video, movies and teleplays, and sports videos. Actually, these three types of video are the most popular and most widely consumed. For example, the authors in [144] use audio features to extract highlights for sports videos. The Broadcast Sportswear Video Retrieval System (SportBB) was developed and proposed by Central China Normal University [118]. Researchers at Taiwan National Chiao Tung University proposed the News Video Browsing System (NVBS) [32]. The News Video Browsing and Retrieval System (NewBR) was researched and developed at Wuhan University [119]. In addition, for movie and teleplay videos, Duminda at George Mason University gives discussion problems and their corresponding solutions using the traditional data mining method for movie video mining and has also designed a film mining system [188]. Meanwhile, most of the researchers have combined video and audio processing technology to mine the hierarchical structure, interested video events, and associations between movies and metadata, as well as to classify movies. Liu and Yana [116] presented a user-centered movie content analysis system. Zhang et al. of Microsoft and the China Academy of Science [216] constructed an emotional movie browsing system through sentiment analysis of movie and teleplay videos. In addition, Lehane et al. [104] used both audio and visual features to detect movie events (important and meaningful video clips).

Movie and teleplay videos are very important resources for television broadcasting. Thus, it is of essence to improve media data's value while providing effective retrieval and a high-resource utilization rate. Although there has been a lot of progress in content-based video retrieval, there are some specific requirements of quick and accurate positioning for video and audio data retrieval for broadcast television. This has dictated that traditional content retrieval cannot be easily obtained through widespread applications in the broadcasting domain. Actually, video and audio data retrieval in broadcast television mainly goes through the following stages: (1) manual retrieval on magnetic tape, (2) manipulator retrieval on magnetic tape, and (3) document-based cataloguing and retrieval. This chapter mainly focuses on document-based cataloguing and retrieval.

In this book, we mainly analyze video cataloguing in the broadcasting domain, that is, deal with movie and teleplay videos. Specifically, we start from the establishment of the association between low-level features and high-level semantics. Unlike textual information, raw video pixel bits do not contain meaningful information by themselves. Therefore, information within a video has to be described through its semantic meaning (i.e., the basic semantic content). Finally, the catalogue file is generated, and based on it, we can realize automatic video cataloguing and efficient retrieval and browsing. As mentioned above, four aspects of videos can be described; we mainly focus on the second and third aspects; video structure data and basic semantic data. Therefore, in this chapter we briefly introduce the research status and progress from these two aspects. More specifically, the key points for video structure parsing are shot boundary detection and video scene detection, while the basic semantic content extraction mainly studies video text extraction and elements of character and locations in the four elements of a whole story (character, time, location, and events).

1.2.1 Related Work on Shot Boundary Detection

The basic idea for shot boundary detection (shot segmentation) is to define frame similarity and detect the obviously changed boundary between shots. After that video structure parsing is conducted by segmenting the whole video into shots. Generally, there are two kinds of shot transitions: abrupt change and gradual change. Abrupt change refers to the last frame of one shot being directly linked to the first frame of another shot without the use of any production effects. Gradual change means that a shot transits to another shot gradually through some special production effects. Gradual transition is always completed in the video postproduction, and there are various types, such as fade in or out, dissolve, and wipe. There has been a large number of literature on shot boundary detection since the 1990s. According to the data forms, all these methods can be divided into two categories: shot boundary detection in the compressed domain and in the noncompressed domain.

Shot boundary detection (SBD) in the compressed domain is shot boundary detection based on the analysis of the DCT (Discrete Cosine Transform) coefficient, the DC (Direct Current) component, the macroblock type, the motion vector characteristics, and so forth. One of the more classic methods was proposed by Yeo [206], analyzing semidecoded DC image sequences. In addition, the DCT coefficient is also usually used for SBD [7, 204]. But for MPEG videos, only I frames can contain the DCT coefficients; thus, the method cannot be used directly in B and P frames, and it always has a high error rate. Therefore, some researchers used the DC coefficient to solve this problem [29], wherein the DC coefficients of the B and P frames are obtained by motion compensation. In addition, literature [90] carries out the analysis to the temporal and spatial distribution of macroblocks, that is, to conduct dissolve transition detection using spatiotemporal distribution of MPEG macroblock types. At the same time, vector quantization techniques have also been applied to shot boundary detection [127]. But this method relies on the video compression ratio; the smaller the compression ratio, the better is the detection performance. In addition to the above methods, there are a lot of SBD methods in H.264/AVC compressed stream. Liu et al. [122] used the intraprediction mode histogram as a computing features and then exploited hidden Markov models (HMMs) to automatically model different cases in which shot transitions can occur among I, P, and B frames. Kim et al. [95] do SBD by analyzing the intra-macroblock partition mode in two successive I frames. In [20], Bita et al. used luminance weighting factors in H.264/AVC to realize the detection of gradual transitions, such as fade.

In spite of shot detection in the compressed domain being very fast, the accuracy of SBD using compression characteristics is not always satisfactory [97]. Better detection accuracy for the shot boundary detection method can often be achieved in the noncompressed domain, for example, methods based on pixels [155], image edges [80], motion vectors [86], and color histograms [130]. Early researchers paid much attentions to abrupt shot boundary detection, and the literature [150, 207] gives a good comparison and analysis of these methods. For gradual transitions, Su et al. [173] utilized the monotonicity of intensity changes during transitions to detect dissolves, and their algorithm can tolerate fast motions. In addition, Yuan et al. [213] proposed a unified shot boundary detection framework based on the graph partition model. Zuzana et al. [230] proposed an efficient detection method using information entropy and mutual information.

At present, abrupt SBD has tended to be mature in the uncompressed domain. As described in [76], these methods solve part of the facing problems in SBD, but do not completely solve the whole problem in SBD, including the efficiency of gradual transition detection and the errors caused by the camera and object motion. Especially, another main challenge is the detection efficiency problem caused by the large data redundancy. In summary, we conducted a simple comparison of the compression domain and uncompressed domain approaches, i.e., approaches in compression domain always have higher efficiency and low accuracy, while lower efficiency but higher accuracy is common for approaches in uncompressed domain.

Compressed domain-based approaches are highly dependent on the compression standards and have the drawback of low reliability and accuracy, especially in the presence of high motion. Although the noncompressed domain-based methods need to deal with more information and relatively low processing efficiency, these shortages have been filled up with the development of computer hardware and software. At the same time, because of that, there is more comprehensive information available in the noncompressed domain, and noncompressed domain-based methods always have more satisfactory detection accuracy. Thus, in this book, we mainly consider the noncompression perspective and study how to get more efficient performance while ensuring detection accuracy.

1.2.2 Related Work on Scene Detection and Recognition

Methods of video scene detection can be roughly divided into two categories: domain dependent and domain independent [219]. Meanwhile, traditional video scene detection methods also include the following types:

1. Scene detection algorithm based on frame difference. These methods take similar ideas in shot boundary detection, that is, consider the position with a large frame difference as a scene boundary [112, 157].
2. Clustering algorithm based on visual features. This type of algorithm clusters shot with the same visual features together; for example, the authors of [114] used k-means clustering, and relaxation iteration is proposed in [140].
3. Shot similarity graph-based methods, such as [152], which presented a similarity metric for shot similarity graph construction and then realized the scene detection by graph segmentation.
4. Other related derivative methods.

Most of the above methods are primarily used as visual features for scene detection. However, visual features do not completely characterize the variation properties in video scenes, especially for movie and teleplay videos. Thus, the multimodality-based methods can more effectively realize scene detection, such as the work in [185].

At the same time, since a video scene is some type of semantic structure unit, the detection of a video scene is somehow related to the semantic content in a specific application. Therefore, in many cases, according to the characteristics of a particular video type, the accuracy of video scene detection can be significantly improved by using the corresponding model with consideration of domain and structure knowledge. Moreover, it can realize the scene classification satisfactorily. At present, there are many methods specifically for movie and teleplay scene detection. Truong et al. [181] use filming and production rules, as well as domain knowledge, to decompose a whole movie into several logical story units, that is, video scenes. The authors in [153] proposed a combination using motion information, shot length, and color features for movie scene detection and achieved good performance. Considering that shots belonging the same scenes have the same background, Chena et al. [35] analyzed the background image of each shot for video scene detection.

While video scene detection divides the whole video into several semantic units, scene recognition mainly answers questions such as "Where it is?" and "Is this the bedroom, street, or office?" That is, it gives an image or video scene a certain semantic label of location and scene category. Actually, most of the scene recognition method is based on a static image database [57, 190]. The realization process generally gives several specific scene categories in advance, and then trains and recognizes the categories of test images through machine learning methods. There are some commonly used image databases for scene recognition, including established by Li et al. a 13-category picture scene database with bedrooms, kitchen, office, high-speed road, and so on, [57]; an 8-category picture scene database built by Oliva and Torralba [143]; and a dataset containing 15 classes of scene images established by Lazebnik et al. [102]. In addition, Xiao et al. [192] established an wider and more meaningful dataset that contains 899 scene categories and 130,519 pictures, named over the Sun database. However, for video scene recognition, as we know, the most related dataset is the Hollywood database, established by Marszalek et al. [129].

Scene recognition is broadly divided into the method based on global features and the method based on local features. The most classic methods that uses global features is the one proposed by Oliva and Torralba [143] with the Gist feature. This method has very good effects in the identification and classification of outdoor scenes. However, while indoor images are involved, their effectiveness was significantly decreased. Actually, compared with the global features, local features have a stronger describing ability and classification characteristics. There are several typical scene recognition methods that use local features, such as a proposed hierarchical model. And also, Lazebnik et al. [102]; Li et al.'s [57] adopted the spatial pyramid matching method, and Wu and Rehg [190] defined the CENTRIST feature descriptor.

However, the above-mentioned methods are proposed to solve the image scene recognition problem and cannot be directly used for video scene classification and recognition with satisfactory results. At present, less work has been proposed for video scene recognition, such as the studies of [84] and [158], and they are mostly for specific simple videos, such as basketball, soccer, and ad videos. Scene classification for movies and teleplays is more challenging because of the diversity and variations of the shooting angle, occlusion, luminance, and so on. So far, the most related works for movie and teleplay scene recognition are in [51] and [50]. In [51], video scene category labeling is realized by the alignment of scripts and video clips. However, the movie and teleplay scripts do not always exist. Therefore, Marszalek et al. [129] proposed a combination of movie action and scene recognition with various local feature extractions.

1.2.3 Related Work on Video Text Recognition

Video text recognition is a very effective and essential processing for video understanding of high-level semantic data extraction. That is, video text can be used as a valuable and direct source for automatic video cataloguing. Here, what we mean by video text is superimposed text in videos, which is distinct for scene text (a part of the background). Generally, video text recognition processing implicitly includes two other preprocessing aspects: video text detection and video text extraction.

Actually, many methods have been proposed for video text detection in the last decade [34, 61, 139], and some of them have good performance. Most existing methods can be roughly classified into three categories: (1) text detection based on texture [134, 205]; (2) text detection based on connected component (CC) [93, 106]; and (3) text detection based on color information [128, 146]. In addition, there are other methods that were not included in the above-mentioned two categories. For example, a hybrid approach is proposed in [145], and the motion perception field (MPF) has also been proposed to locate text regions especially with continuous video frames, as introduced in our previous work [85].

Meanwhile, the existing video text extraction methods can be roughly classified into four groups: threshold-based methods [67], stroke-based methods [109], cluster-based methods [87, 96], and others [147, 148].

1.2.4 Related Work on Character Identification

The task of associating faces with names in a movie or TV program is typically accomplished by combining multiple sources of information, for example, image, video, and text, with little or even no manual intervention. In the early stages, the most similar application is to identify faces in news videos [18, 33, 163], especially recognizing the announcers or specific characters (politicians or star actors). In news videos, the labeled names are always available in captions or transcripts, and the appearances of these people are also very clear and distinct. However, in movies or TV series, the names of characters are not always available, and the appearances of characters vary in different conditions, which makes it hard to detect, track, and recognize these characters.

Over the past two decades, extensive research efforts have actively concentrated on this task in movies and TV series. Since we need to assign the character names to faces or bodies in videos, the set of names is necessary in advance. According to the utilized contents or clues for these names, previous work can be roughly classified into two groups:

Group 1 studies the labeling task for character recognition utilizing manually labeled visual or audio data as the training dataset. In this group, supervised learning or semisupervised learning methods are used; namely, the researchers collect several samples as training data to generate the recognition model, and the labeling information is used as the final labeling text. Arandjelovic and Zisserman [9] used face images as a query to retrieve particular characters. Affine warping and illumination correcting were utilized to alleviate the effects of pose and illumination variations. In the work of Liu and Wang [123], a multicue approach combining facial features and speaker voice models was proposed for major cast detection. In our previous work [195], we also proposed a semisupervised learning strategy to address celebrity identification with collected celebrity data. In addition, several other methods have used audio clues or both audio and vision clues [91, 99, 110]. However, these approaches cannot automatically assign real names to the characters. Therefore, most of the researchers use the manually labeled training data. For example, Tapaswi et al. [178] presented a probabilistic method for identifying characters in TV series or movies, and the face and speaker models were trained on episodes 4–6 with manual labeling. Although there are many seasons and episodes in a TV series, there might be insufficient data used for training for movies, which have only one episode with a duration of about 2 h.

Group 2 handles the problem of assigning real names to the characters by using textual sources, such as scripts and captions [54, 99]. Guillaumin et al. [70] presented methods for face recognition using a collection of images with captions, especially for news video. Everingham et al. [54] proposed employing readily available textual sources, the film script and subtitles, for text video alignment, and thus obtained certain annotated face exemplars. The rest of the faces were then classified into these annotated exemplars. Their approach was followed by Laptev et al. [231] for human action annotation. However, in [54], the subtitle text and time stamps were extracted by optical character recognition (OCR), which required extra computation cost on spelling error correction and text verification. In addition, Zhang et al. [218] investigated the problem of identifying characters in feature-length films using video and film scripts with global face–name matching. Bojanowski et al. [23] learned a joint model of actors and actions in movies using weak supervision-provided scripts. In fact, except for the extra errors and computation cost, for some movies, the scripts cannot be found easily or may be quite different from subtitles.

In addition, the above-mentioned methods always take each face image or face track for recognition individually. They are strictly limited to the characters' face appearance, including face size, angle, and resolution. Further, since the identification is performed for individual tracks, constraints such as the same person cannot appear twice in one frame and faces in continuous shots may with small probability be the same character cannot be integrated. Thus, some methods have taken into consideration these constraints with the probability-generated models [8, 13, 99]. For example, Anguelov et al. [8] proposed a method to recognize faces using the markov random field (MRF) model on photo albums, which performed recognition primarily based on faces and incorporated clothing features from a region below the face.

1.3 Main Research Work

In this book, we discuss and explore research on video cataloguing, especially movie and teleplay cataloguing, from two aspects: (1) video structure parsing, including shot boundary detection, key frame extraction, and video scene detection, and (2) basic semantic word extraction, including video text extraction, video scene recognition, and movie character identification.

1. **Shot boundary detection**. The transformation of color and texture appearances in movies and teleplays is not as obvious as that in news video and advertising video. Hence, most traditional shot boundary detection methods do not perform well in movie or teleplay shot boundary detection. Facing the two major issues for shot boundary detection—how to effectively describe the interframe difference for accurate detection and how to reduce redundancy information processing for rapid implementation—we focus on the analysis and study of the combination of accuracy and efficiency. Finally, an accelerating SBD approach is presented using mutual information calculation on a focus region and skipping interval. In fact, we not only accelerated the processing in both the spatial and temporal domains, but also archived a satisfactory accuracy. In addition, we used corner distribution analysis to remove the false detections based on the fact that different candidate shot boundaries (i.e., abrupt shot boundary, gradual transition, and motion-caused false detection) show different corner distributions.
2. **Video scene segmentation and recognition**. Scene backgrounds are complex and have wide variety of changes. Thus, we combined the characteristics of audio and video features. That is, we extracted visual features (including color, low edge, and motion information) and audio features (including mel-frequency cepstral coefficients, short-time energy, and zero crossing rate), and then combined them with their correlations using the kernel-based canonical correlation analysis method. After that, shot similarity graph partitioning is involved getting all the video scenes. For these video scenes, video scene

recognition is adopted to give a location label to each video scene. In view of the characteristics, video scene background takes less percentage of a frame and the appearance changes quickly. However, there are several inherent characteristics, such as the redundancy of key frames covering a wide range content and the human regions always shielding the appearance of a scene. Hence, we combined the advantages of mature image scene recognition methods and these inherent characteristics to extract the representative local features, and then we used the bag of words and model theme to do scene recognition model training. At the same time, considering the repetition of video scenes, similar correlations among video scenes, were constructed and used for enhancing scene identification.

3. **Video text recognition**. Video text often contains plentiful semantic information that is significant and crucial for video retrieval and cataloguing. Generally, video text can be classified into superimposed text and scene text. Superimposed texts are added by the video editor in postediting, and these texts are the condensed semantic description of the video clips. Scene text is inherent in the video captured by the video camera and is always an integral component of the background. Generally, scene text detection and recognition is very important in surveillance video analysis and the computer vision area, and superimposed text detection and recognition is always used for broadcast video analysis. While superimposed text is a sort of human-summarized semantic content with condensed meaning, it is perfect for video summary and cataloguing. Hence, in this book, we only pay attention to superimposed detection and recognition. Thus, in this book, video text refers to superimposed text.

4. **Character identification**. Generally, characters and their actions are the most crucial clues in movies and TV series. Therefore, face detection, character identification, human tracking, and so on, become important in content-based movie and TV series analysis and retrieval. Meanwhile, the characters are often the most important contents to be indexed in a feature-length movie; thus, character identification becomes a critical step in film semantic analysis, and it is also reasonable to use character names as cataloguing items. This book mainly describes how to recognize characters in movies and TV series and then use these character names as cataloguing items for an intelligent catalogue.

Besides, in order to verify the feasibility of the research results, a prototype system of a automatic movie and teleplay cataloguing (AMTC) is designed and implemented based on the above-introduced methods. Meanwhile, except for movies and teleplays, we also introduced other related applications, such as highlight extraction in basketball.

Chapter 2

Visual Features Extraction

Features extraction is the very important primary process for almost all video analysis and video processing applications. The good and bad of feature extraction directly influence the following processes, and also determine the final performance of video analysis. Over the last years, many feature extraction techniques have been proposed, and most of them are robust and efficient in some specific domain. In this chapter, we list and introduce some of the popularly used features in video analysis, especially visual features.

2.1 Introduction

The definition of feature discussed here is a concept in computer vision and video processing where a feature is a piece of information that is the abstraction of the input data and relevant for solving computational tasks. Meanwhile, feature extraction is the process or method to extract features from input data (video in this book), and the definition of it in Wikipedia is

> *In pattern recognition and in image processing, feature extraction is a special form of dimensional reduction. When the input data to an algorithm is too large to be processed and it is suspected to be notoriously redundant (e.g., the same measurement in both feet and meters) then the input data will be transformed into a reduced representation set of features (also named features vector). Transforming the input data into the set of features is called feature extraction. If the features extracted are carefully chosen it is expected that the features set will extract the relevant information from the input data in order to perform the desired task using this reduced representation instead of the full size input.*

Feature extraction is the key step for video processing. How to extract ideal features that can reflect the intrinsic content of the videos as completely as possible is still a challenging problem. As we know, the basic visual feature contains color, texture, shape, and so on. Generally, visual features can always be classified into global and local features. However, here, what we discuss is some typical or classical features used especially in video processing and the computer vision domain. The rest of this chapter describes them one by one. After that, a novel idea for feature extraction, named feature learning, is also introduced.

2.2 Scale-Invariant Feature Transform

Scale-invariant feature transform (SIFT) is nearly the most popular local feature of the past 10 years; it was published by David Lowe in 1999 [124, 125]. Lowe's method for image feature generation transforms an image into a large collection of feature vectors, each of which is invariant to image translation, scaling, and rotation; partially invariant to illumination changes; and robust to local geometric distortion.

SIFT is invariant to image scale and rotation and robust to affine distortion, change in three-dimensional (3D) viewpoint, addition of noise, and change in illumination. As described in [125], there are four major stages in getting a set of image features: scale-space extrema detection, key point localization, orientation assignment, and key point descriptor.

1. **Detection of scale-space extrema**. This stage is used to identify locations and scales that can be repeatably assigned under differing views of the same object. Specifically, the scale space of an image is defined as a function $\mathcal{L}(x, y, \sigma)$ that is produced from the convolution of a variable-scale Gaussian, $G(x, y\sigma)$, to the input image $I(x, y)$:

$$\mathcal{L}(x, y, \sigma) = G(x, y, \sigma) * I(x, y) \tag{2.1}$$

$$G(x, y, \sigma) = \frac{1}{2\pi\sigma^2} e^{-(x^2+y^2)/2\sigma^2} \tag{2.2}$$

where $*$ means the convolution operation in x and y. Then, the scale-space extrema in the difference-of-Gaussian (DoG) function convolved with the image is calculated from the difference of two nearby scales separated by a constant multiplicative factor k:

$$D(x, y, \sigma) = (G(x, y, k\sigma) - G(x, y, \sigma)) * I(x, y)$$

$$= \mathcal{L}(x, y, k\sigma) - \mathcal{L}(x, y, \sigma) \tag{2.3}$$

The efficient approach to construct $D(x, y, \sigma)$ is shown in Figure 2.1.

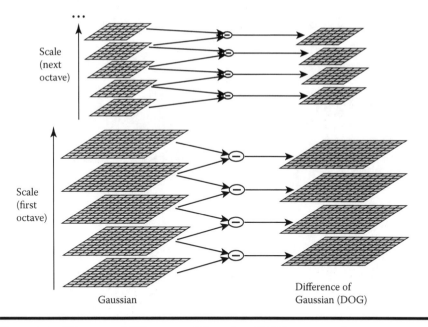

Gaussian

Difference of
Gaussian (DOG)

Figure 2.1   Diagram of difference-of-Gaussian (DOG) construction.

2. **Key point localization.** In order to localize the local maxima and minima extreme point of $D(x, y, \sigma)$, each sample point is compared with its neighbors in the image and scale domains, as shown in Figure 2.2, including eight neighbors in the current image and eight in the scale above and below. It is selected only as a key point candidate if it is larger than all of these neighbors or smaller than all of them. The cost of this check is reasonably low due to the fact that most sample points will be eliminated following the first few checks.

So far, those key point candidates that have low contrast (and are therefore sensitive to noise) or are poorly localized along an edge should to be rejected. Taylor expansion (up to the quadratic terms) of the scale-space function,

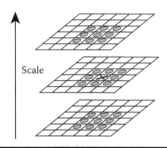

Scale

Figure 2.2   Sample points and their neighbors.

$D(x, y,)$, shifted so that the origin is at the sample point:

$$D(x) = D + \frac{\partial D^T}{\partial \mathbf{x}} \mathbf{x} + \frac{1}{2} \mathbf{x}^T \frac{\partial D^T}{\partial \mathbf{x}^2} \mathbf{x} \tag{2.4}$$

where $\mathbf{x} = (x, y, \sigma)^T$.

$$D(\hat{\mathbf{x}}) = D + \frac{1}{2} \frac{\partial D^T}{\partial \mathbf{x}} \hat{\mathbf{x}} \tag{2.5}$$

where $\hat{\mathbf{x}}$ is defined as

$$\hat{\mathbf{x}} = -\frac{\partial^2 D^{-1}}{\partial \mathbf{x}^2} \frac{\partial D}{\partial \mathbf{x}} \tag{2.6}$$

Thus,

$$D(\hat{\mathbf{x}}) = D + \frac{1}{2} \frac{\partial D^T}{\partial \mathbf{x}} \hat{\mathbf{x}} \tag{2.7}$$

The extrema with a value of $D(\hat{\mathbf{x}})$ less than 0.03 were discarded. In addition, a poorly defined task in the difference-of-Gaussian function will have a large principal curvature across the edge but a small one in the perpendicular direction, where the principal curvature is computed from a 2×2 Hessian matrix H,

$$\begin{bmatrix} D_{xx} & D_{xy} \\ D_{xy} & D_{yy} \end{bmatrix} \tag{2.8}$$

Actually, we only need to check if the following equation is satisfied:

$$\frac{Tr(H)^2}{Det(H)} < \frac{(r+1)^2}{r} \tag{2.9}$$

3. **Orientation assignment.** For each image sample, $\mathcal{L}(x, y)$, at this scale, the gradient magnitude, $m(x, y)$, and orientation, $\theta(x, y)$, are computed using pixel differences:

$$m(x, y) = \sqrt{(\mathcal{L}(x+1, y) - \mathcal{L}(x-1, y))^2 + (\mathcal{L}(x, y+1) - \mathcal{L}(x, y-1))^2} \tag{2.10}$$

$$\theta(x, y) = \alpha \tan 2((\mathcal{L}(x, y+1) - \mathcal{L}(x, y-1))(\mathcal{L}(x+1, y) - \mathcal{L}(x-1), y)) \tag{2.11}$$

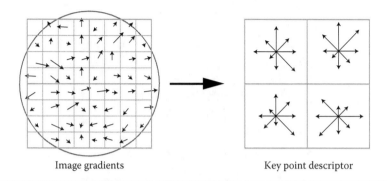

| Image gradients | Key point descriptor |

Figure 2.3   **The construction of a SIFT descriptor.**

The orientation histogram has 36 bins covering the 360° range of orientations. Each sample added to the histogram is weighted by its gradient magnitude and by a Gaussian-weighted circular window with a σ that is 1.5 times that of the scale of the key point.

4. **Key point descriptor.** A key point descriptor is created by first computing the gradient magnitude and orientation at each image sample point in a region around the key point location, as shown on the left in Figure 2.3. These are weighted by a Gaussian window, indicated by the overlaid circle. These samples are then accumulated into orientation histograms summarizing the contents over 4×4 subregions, as shown on the right, with the length of each arrow corresponding to the sum of the gradient magnitudes near that direction within the region. This figure shows a 2×2 descriptor array computed from an 8×8 set of samples, whereas the experiments in this chapter use 4×4 descriptors computed from a 16×16 sample array. Generally, we use a $4 \times 4 \times 8 = 128$ descriptor.

2.3 Gabor Feature

A Gabor filter, named after Dennis Gabor, is a linear filter used for edge detection. Frequency and orientation representations of Gabor filters are similar to those of the human visual system, and they have been found to be particularly appropriate for texture representation and discrimination. In the spatial domain, a two-dimensional (2D) Gabor filter is a Gaussian kernel function modulated by a sinusoidal plane wave.[1]

A set of Gabor filters with different frequencies and orientations may be helpful for extracting useful features from an image. In image and video analysis areas, the

---

[1] http://en.wikipedia.org/wiki/Gabor_filter.

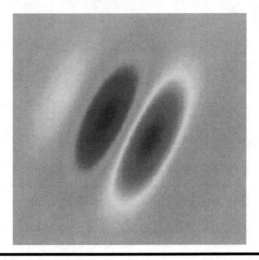

Figure 2.4   Example of a two-dimensional Gabor filter.

frequently used one is the 2D Gabor filter (Figure 2.4). The core of Gabor filter–based feature extraction is the following 2D Gabor filter function:

$$\psi(x, y) = \frac{f^2}{\pi \gamma \eta} e^{-\left(\frac{f^2}{\gamma^2}(x')^2 + \frac{f^2}{\eta^2}(y')^2\right)} e^{j2\pi f x'} \tag{2.12}$$

$$x' = x \cos\theta + y \sin\theta \tag{2.13}$$

$$y' = x \sin\theta + y \cos\theta \tag{2.14}$$

where f is the central frequency of the filter, θ is the rotation angle of the Gaussian major axis and the plane ware, γ is the sharpness along the major axis, and η is the sharpness along the minor axis. In the given form, the aspect ratio of the Gaussian $\lambda = \frac{\eta}{\gamma}$. Then, the normalized 2D Gabor filter function in the frequency domain is defined as

$$\Psi(u, v) = e^{-\frac{\pi^2}{f^2}(\gamma^2(u'-f)^2 + \eta^2(v')^2)} \tag{2.15}$$

$$u' = u \cos\theta + v \sin\theta, \quad v' = -u \sin\theta + v \cos\theta \tag{2.16}$$

The effects of the parameters, interpretable via the Fourier similarity theorem, are demonstrated in Figure 2.5.

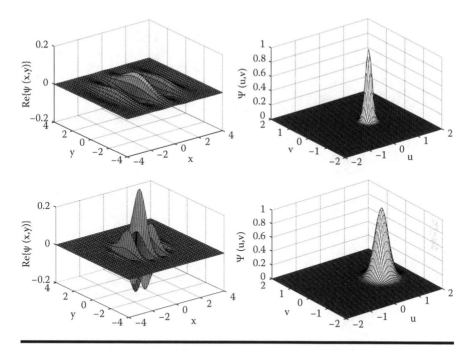

Figure 2.5 Examples of 2D Gabor filters in the spatial and frequency domains: (a) $f = 0.5$, $\theta = 0$, $\lambda = 1.0$, $\eta = 1.0$; **(b)** $f = 1.0$, $\theta = 0$, $\lambda = 1.0$, $\eta = 1.0$.

2.4 Histogram of Oriented Gradients (HOG)

Generally speaking, histograms of oriented gradients (HOGs) are feature descriptors used in computer vision and image processing for the purpose of object detection. The technique counts occurrences of gradient orientation in localized portions of an image. This method is similar to that of edge orientation histograms, scale-invariant feature transform descriptors, and shape contexts, but it differs in that it is computed on a dense grid of uniformly spaced cells and uses overlapping local contrast normalization for improved accuracy.

Navneet Dalal and Bill Triggs, researchers for the French National Institute for Research in Computer Science and Control (INRIA), first described histogram of oriented gradient descriptors in their June 2005 Computer Vision and Pattern Recognition (CVPR) paper [2]. In this work they focused their algorithm on the problem of pedestrian detection in static images, although since then they have expanded their tests to include human detection in film and video, as well as a variety of common animals and vehicles in static imagery.

According to Dalal and Triggs's description, the HOG descriptor has the following properties: RGB color space with no gamma correction, [−1, 0, 1] gradient filter with no smoothing, linear gradient voting into nine orientation bins in 0°–180°,

Figure 2.6 Examples of MSER regions.

16×16 pixel blocks of four 8×8 pixel cells, Gaussian spatial window with $\sigma = 8$ pixel, $L2 - Hys$ (Lowe-style clipped $L2$ norm) block normalization, block spacing stride of 8 pixels, 64×128 detection window, and linear support vector machine (SVM) classifier.

2.5 Maximally Stable Extremal Regions

Maximally stable extremal regions (MSERs) were proposed by Matas et al. [1] to find correspondences between image elements from two images with different viewpoints; that is, MSER regions are connected areas characterized by almost uniform intensity, surrounded by contrasting background (Figure 2.6).

With the definitions in literature [1], the computation of MSER can be shown as follows:

Image: A mapping $I : \mathcal{D} \subset \mathbb{Z}^2 \rightarrow \mathcal{S}$. Extremely regions are well defined on images if

1. $\mathcal{S}$ is totally ordered; that is, a reflexive, antisymmetric, and transitive binary relation $\leq$ exists. In this chapter only $\mathcal{S} = 0, 1, \dots, 255$ is considered, but extremal regions can be defined on, e.g., real-valued images ($\mathcal{S} = R$).
2. An adjacency (neighborhood) relation $A \subset \mathcal{D} \times \mathcal{D}$ is defined. In this chapter, four-neighborhoods are used; that is, $p, q \in \mathcal{D}$ are adjacent ($p \, A q$) if and only if $\sum_{i=1}^{d} |p_i - q_i| \leq 1$.

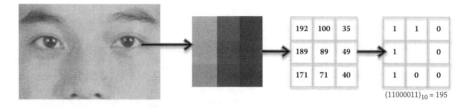

$(11000011)_{10} = 195$

Figure 2.7   Illustration of the basic LBP operator.

Region: $\mathcal{Q}$ is a contiguous subset of $\mathcal{D}$; that is, for each $p, q \in \mathcal{Q}$ there is a sequence $p, a_1, a_2, \ldots, a_n, q$ and $p A a_1, a_i A a_{i+1}, a_n A q$.

(Outer) region boundary: $\partial \mathcal{Q} = \{p \in \mathcal{D} \backslash \mathcal{Q} : \exists p \in \mathcal{Q} : q A p\}$; that is, the boundary $\partial \mathcal{Q}$ of $\mathcal{Q}$ is the set of pixels adjacent to at least one pixel of $\mathcal{Q}$ but not belonging to $\mathcal{Q}$.

Extremal region: $\mathcal{Q} \subset \mathcal{D}$ is a region such that for all $p \in \mathcal{Q}, q \in \partial \mathcal{Q}$: $I(p) > I(q)$ (maximum intensity region) or $I(p) < I(q)$ (minimum intensity region).

Maximafly stable extremal region: Let $\mathcal{Q}_1, \ldots, \mathcal{Q}_{i-1}, \mathcal{Q}_i, \ldots$ be a sequence of nested extremal regions, that is, $\mathcal{Q}_i \subset \mathcal{Q}_{i+1}$. Extremal region $\mathcal{Q}_{i*}$ is maximally stable if and only if $q(i) = |\mathcal{Q}_{i+\triangle} \backslash \mathcal{Q}_{i-\triangle}| // |\mathcal{Q}_i|$ has a local minimum at i^* ($| \bullet |$ denotes cardinality). $\triangle \subset \mathcal{S}$ is a parameter of the method. The equation checks for regions that remain stable over a certain number of thresholds. If a region $\mathcal{Q}_{i+\triangle}$ is not significantly larger than a region $\mathcal{Q}_{i-\triangle}$, region $\mathcal{Q}_i$ is taken as a maximally stable region.

2.6 Local Binary Pattern (LBP)

Local binary patterns (LBPs) are a type of feature used for classification in computer vision. LBP is a particular case of the texture spectrum model proposed in 1990 [78,79]. The first method using local binary patterns was proposed by Ojala et al. [141]. Meanwhile, the original LBP operator is defined in a 3×3 window, which centers on the center pixel. The center pixel is compared with its eight neighbors, and if the value of its neighbor is bigger, the corresponding label is set to 1, otherwise 0. As the neighborhood consists of eight pixels, a total of $2^8 = 256$ different labels can be obtained depending on the relative gray values of the center and the pixels in the neighborhood. Figure 2.7 shows an illustration of the basic LBP operator.

Generally, we can get the LBP feature vector with the following processes:

■ Divide the examined window into cells (e.g., 16×16 pixels for each cell).
■ For each pixel in a cell, compare the pixel to each of its eight neighbors (on its left top, left middle, left bottom, right top, etc.). Follow the pixels along a circle, that is, clockwise or counterclockwise.

- Where the center pixel's value is greater than the neighbor's value, write 1. Otherwise, write 0. This gives an eight-digit binary number (which is usually converted to decimal form for convenience).
- Compute the histogram, over the cell, of the frequency of each number occurring (i.e., each combination of which pixels are smaller and which are greater than the center).
- Optionally normalize the histogram.
- Concatenate (normalize) histograms of all cells. This gives the feature vector for the window.

2.7 Feature Learning

Much recent work in machine learning has focused on learning good feature representations from unlabeled input data for higher-level tasks such as classification. Meanwhile, feature learning or representation learning is a set of techniques to learn a transformation of raw inputs to a representation. Developing domain-specific features for each task is expensive, time-consuming, and requires expertise of the data. The alternative is to use unsupervised feature learning [16, 39, 53] in order to learn a layer of feature representations from unlabeled data. The advantage of learning features from unlabeled data is that the plentiful unlabeled data can be utilized and potentially better features than handcrafted features can be learned. Both of these advantages reduce the need for expertise of the data. Actually, deep networks have been used to achieve state-of-the-art performance on many benchmark datasets and for solving challengeable artificial intelligence (AI) tasks. More details of a typical feature learning are shown in [39].

In [39], the random subpatches are extracted from unlabeled input images at first, and each patch has dimension $\omega \times \omega$ and d channels (ω is referred to as the receptive field size). Each $\omega \times \omega$ patch can be represented as a vector in $\mathbb{R}^N$ of pixel intensity values, with $N = w \cdot w \cdot d$, and then a dataset of m randomly sampled patches, $X = x^{(1)}, \ldots, x^{(m)}$, where $x^{(i)} \in \mathbb{R}^N$.

Given the dataset, each patch $x^{(i)}$ is normalized by subtracting the mean and dividing by the standard deviation of its elements. For visual data, this corresponds to local brightness and contrast normalization. After that, the entire dataset X may optionally be whitened [88].

Generally, an unsupervised learning algorithm, or the feature learning, is looked at as a black box that takes the dataset X and outputs a function $f : \mathbb{R}^N \to \mathbb{R}^K$. This function refers mapping each path $x^{(i)}$ to a new feature vector of K features. Actually, there are several typically used unsupervised learning algorithms to do feature learning on the dataset.

Sparse autoencoder. An autoencoder with K hidden nodes using backpropagation is used to minimize squared reconstruction error with an additional penalty term that encourages the units to maintain a low average activation [103]. The output weights $W \in \mathbb{R}^{K \times N}$ and biases $b \in \mathbb{R}^K$ so that the feature mapping f is

$$f(x) = g(Wx + b) \qquad (2.17)$$

where $g(z) = 1/(1 + \exp(-z))$ is the logistic sigmoid function.

Sparse restricted Boltzmann machine. A restricted Boltzmann machine (RBM) is a generative model. The RBM is likely the most popular subclass of Boltzmann machine [171]. The RBM can be said to form a bipartite graph with the visibles and the hidden forming two layers of vertices in the graph (and no connection between units of the same layer) with K binary hidden variables. Sparse RBMs can be trained using the contrasting divergence approximation [82] with the same type of sparsity penalty as the autoencoders.

Gaussian mixtures. Gaussian mixture models (GMMs) represent the density of input data as a mixture of K Gaussian distributions and are widely used for clustering. GMMs can be trained using the expectation–maximization (EM) algorithm, as in [5]. We run a single iteration of K-means to initialize the mixture model. The feature mapping f maps each input to the posterior membership probabilities:

$$f_k(x) = \frac{1}{(2\pi)^{d/2} |\sum_k|^{1/2}} \cdot \exp\left(-\frac{1}{2}\left(x - c^{(k)}\right)^T \sum_k^{-1} \left(x - c^{(k)}\right)\right) \qquad (2.18)$$

where $\sum_k$ is a diagonal covariance and ϕ_k are the cluster prior probabilities learned by the EM algorithm.

With the above-mentioned unsupervised learning algorithms, a function f refers to a mapping from input patch $x \in \mathbb{R}^N$ to a new representation $y = f(x) \in \mathbb{R}^K$, and this is what is meant by unsupervised feature learning.

2.8 Summary

Feature extraction is to build derived values (features) intended to be informative, nonredundant, and facilitate the subsequent learning and generalization steps, in some cases leading to better human interpretations. Meanwhile, feature extraction is the primary processing for almost all of the video analysis and video processing applications. As we know, the basic visual feature contains color, texture, shape, and so on. Generally, visual features can always be classified into global and local features. Actually, in this chapter, we discussed some typical or classical features especially used in video processing and the computer vision domain, including scale-invariant feature transform (SIFT), Gabor features, histogram of gradients (HOG), maximally stable extremal regions (MSERs), and local binary pattern (LBP). Finally, a novel idea for feature extraction, feature learning, was also introduced briefly.

Chapter 3

Accelerating Shot Boundary Detection

In this chapter, we introduce and discuss the most popular and recent shot boundary detection methods. Then, we analyze the accuracy and efficiency of shot boundary detection. Finally, a novel accelerated shot boundary detection approach is proposed and compared with existing methods to verify its efficiency and accuracy.

3.1 Introduction

Video processing and analysis has become more and more important since the rapid emergence of video resources with increased availability of video cameras. Video segmentation is one of the fundamental steps in video processing and analysis. Generally, videos are segmented using shot boundary detection (SBD) as the basic approach. Here, a shot is defined as a continuous sequence of frames captured using one camera [26]. Usually, all the frames in a shot have consistent visual characteristics, such as color, texture, and motion.

The transitions between two shots can be abrupt or gradual, and these transitions can be classified into CUT transitions or gradual transitions (GTs) respectively [41]. A CUT transition is an instantaneous transition from one shot to the next. There are no transitional frames between two shots. The gradual transition is formed by the editor to insert an effect of a fade, wipe, or dissolve so that the frames change slowly within a span of several frames. For GT, the transitions are realized little by little which will last for dozens of frames. Fade in or out or dissolve is the most frequently used GT type. A dissolve is a gradual transition from one image to another. The terms *fade in* and *fade out* are used to describe a transition to and from a blank image. A dissolve overlaps two shots for the duration of the effect, usually at the end of one

Figure 3.1   An example of a gradual transition shot boundary in news video.

shot and the beginning of the next, but may be used in montage sequences also. Figure 3.1 shows a gradual transition shot boundary example in a news video.

The change of shot is the variation of scene content. Because of the various descriptions for scene content, over the last decade, there have been many approaches proposed for shot boundary detection [4, 80, 155, 179, 207, 227]. Meanwhile, the main research area for shot boundary detection includes scene content representation, feature extraction, robustness, or efficiency of the method. For the present, shot boundary detection has made a lot of progress; a lot of algorithms have been proposed and been applied to content-based retrieval systems. However, the related theories, including feature extraction and content presentation, are not so robust or perfect. There are still several problems for shot boundary detection [73, 74]: (1) how to extract the suitable features, (2) motions and noises in video, (3) false negatives for GT detection, (4) how to choose an optimal threshold, and (5) computational complexity.

Focusing on the above-mentioned problems, we discuss how to decrease the computational complexity along with other problems. In particular, we emphatically study:

1. How to reduce the redundant information and efficiently describe the difference between frames.
2. How to accurately detect nearly all shot boundaries as well as remove false alarms, especially for GT boundaries.
3. How to complete the shot boundary detection as quickly as possible.

Actually, since the large computing cost is caused by the computation of large-scale frames and pixels, we discuss how to reasonably reduce the processed frames and pixels to promote efficiency, as well as extract more robust features to ensure accuracy. In addition, considering that GTs are harder to detect than cuts, we introduce an adaptive skipping detection algorithm to more accurately find almost all GTs, regardless of fade in or out, wipe, and so on.

The main objectives of this chapter are to:

1. Summarize the related works in shot boundary detection.
2. Propose a method to describe differences between frames with consideration of focus regions' (FRs) mutual information, which successfully reduces the processed pixels in the spatial domain.
3. Accelerate the SBD process in the temporal domain by skipping frames. Except for exponentially increasing the detection speed, almost all boundaries could be detected, including GTs, which are hard to detect.
4. Explain skipping detection, which involves several false alarms, especially false detections caused by object and camera motion.
5. Conduct extensive experiments to show that our approach can not only speed up SBD, but also detect shot boundaries with high accuracy in both CUT and GT boundaries.

3.2 Related Work

The basic idea for all shot boundary detections is to analyze the video frame sequence to find the difference between neighbor frames, and consequently, to determine if the frames belong to the same shot. Cotsaces et al. [41] gives a comparison for most of the recent methods. Actually, the most used SBD methods can be classified into the following groups:

1. **Gray value–based SBD methods**. Methods in this group directly compute the gray difference between frames. Specifically, gray differences of pixels in the corresponding positions are computed and then summated together to form the total difference. Then a threshold is used to judge if the difference of frame means a shot boundary. Because the gray value is a direct description of color and light in the strict corresponding position, these methods are always sensitive to changes in light, camera, moving objects, and so on.
2. **Edge-based SBD methods**. The points at which image brightness changes sharply, or more formally, have discontinuities that are typically organized into a set of curved line segments call edges. Image edges can reflect the graph and object contours well; thus, they are a very popular feature in image processing. Generally, methods based on edges [80, 211] always use the Sobel operator, Canny operator, or Laplace operator to get the edge maps. By analyzing the changes of the edge pixels to determine the shot boundary, Yoo et al. [211] adopted the Sobel operator to get the average edge distribution framework with information fusion and offset. And then, by comparing the local edge distribution in each frame with the average distribution framework, they obtained the variation and edge distribution change curve.

Finally, with mathematics analysis on the change curve, the shot boundary was detected.

3. **Color histogram-based SBD methods**. For video or digital images, the intricate scattering of different color is the most intuitive description. Therefore, an intuitive idea is to compare the color histograms of neighbor frames to get shot boundaries [75, 130]. The simplest way is to calculate the histogram difference and then use a threshold to determine if there are any boundaries. Yeung and Liu [209] also found that the normalized color histogram is a better metric of frame difference, and this metric is defined as

$$D(f_i, f_j) = \frac{\sum_b |h_{ib} - h_{jb}|}{N} \tag{3.1}$$

where $h_i b$ is the given gray value of the $i_t h$ frame and N is the total number of pixels in a frame. Besides, based on the statistic of gray value and color histogram, Lalor and Zhang [101] introduced the chi-square measure to quantify the frame difference. In fact, for these methods, the detection accuracy always can be improved with various mathematics or algorithm preprocessing on the histogram. However, the selection of a preprocessing algorithm, as well as the possibility that two different frames may have the same color distribution and color histogram, always causes these methods to fail.

4. **Block matching and motion vector–based SBD methods**. Motion vector is very important for shot boundary detection. Zhang et al. [215] used the motion vector from block matching to determine if the shot boundary is drawing or swapping. Akutsu et al. [6] used a motion smooth metric to detect the shot changes. In this method, first, each frame is segmented into 88 blocks and the corresponding blocks in adjacent frames are matched to calculate the correlation coefficient of optimal matching blocks. Second, the average of these correlation coefficients is used as a similarity metric for SBD.

Nevertheless, most of the above-mentioned methods still cannot completely solve most of the detection problems [76] and only help to alleviate the situation. Besides, Yuan et al. [213] proposed a unified SBD system based on a graph partition model to do formal study of SBD. Huang et al. [83] implemented SBD via local key point matching, and the method based on information theory was also discussed in [191, 230]. However, as described in [76], these methods alleviated the situation to some extent, rather than completely solved the problems caused by camera motion and change of illumination.

Some other works concentrated on the computational complexity since it is another important factor in SBD, especially in real-time applications. For example, Lefévre et al. [166] depicted the computational complexity in different SBD methods with real-time situations. Although some methods were proposed as fast SBD methods, such as [115], they did not really accelerate the processing. Accordingly, the authors of [45] and [111] presented fast SBD methods by skipping frames.

For instance, Danisman and Alpkocak [45] set the interval threshold (named TSFI) as 1 and 5. This cannot efficiently improve the efficiency because if the GTs last for more than five frames, they cannot be accurately found by this small interval. Li et al. [111] used bisection-based comparisons to eliminate nonboundary frames. However, they employed the pixel-wise distance of the luminance component to compare two frames, that are sensitive to color, flush, and so on. Furthermore, they used 20 frames as the skip interval for distance calculation, which reduces the processed frames in a limited scope.

3.3 Frame Difference Calculation

This section discusses how to reduce the processed pixels in a frame, namely, by accelerating the shot boundary detection in the spatial domain. Meanwhile, it also depicts how to combine mutual information and color histograms to efficiently get frame difference measurements in defined focus regions.

3.3.1 Focus Region

A video has thousands of frames, and each frame has thousands of pixels. These vast frames and pixels make the computation complexity very high, which is the reason that many SBD methods or systems are of low efficiency. Although spatial subsampling of frames has been suggested to improve video processing efficiency in [37, 196], it still depends on the choice of the spatial window. A smaller window size is sensitive to object and camera motions, while an arbitrary window size does not allow the remaining pixels to represent the frame well. For example, the authors of [37] only used the histogram difference of upper halves in two successive frames, which generated many false boundaries because of the loss of important information in the lower halves.

Generally, the most essential information in a frame is concentrated around the center of a frame, and the more the pixels are close to the frame center, the more important they are. In order to reduce the cost time, redundant pixels should be removed and only informative pixels kept for processing. To accomplish this, a focus region (FR) is defined for each frame, as shown in Figure 3.2. The FR of a $P \times Q$–sized frame is extracted in the following steps:

1. Each image is divided into nonoverlapping regions of size $(P/p) \times (Q/q)$ to get $p \times q$ number of subregions.
2. The first external round subregions (colored red in Figure 3.2) are defined as the nonfocus region.
3. Then, we take the second external round subregions (yellow circle subregions) as the second focus region.
4. What remains in the center is the most focused region.

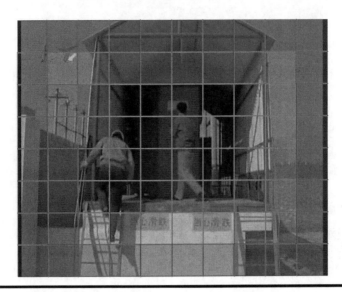

Figure 3.2   Illustration of an FR. The frame is partitioned into 8 × 10 subregions, and the outermost round subregions are nonfocus regions, the second outermost round subregions are the second focus regions, and the remaining subregions are the most focused regions.

In order to efficiently represent the frame with the chosen subset of pixels that exclude the unimportant pixels, we remove pixels in the nonfocus region, fully use pixels in the most focused region, and select pixels with odd x-coordinates in the second focus region. At last, we can get a new set of pixels to present the frame, named FR, and only pixels in FR are used for the next processing.

3.3.2 Frame Difference Measurement

Mutual information (MI) [42] is an important concept in information theory to measure the amount of information transported from one random variable to another. It measures the reduction of uncertainty of one random variable given the knowledge of another. Actually, the authors of [191, 230] utilized the MI between two video frames to measure the difference for SBD, and these methods achieved satisfactory performance. Thus, the MI is used in our approach for frame similarity measure.

Although most of the recent SBD methods can do well in CUT detection, several of them are sensitive to GT detection because the changes between two consecutive frames in GT are not so evident (shown in Figure 3.3). While we can observe that the consecutive frames are similar in GT, we cannot get the difference of them by simple

Figure 3.3 **Corner distribution in gradual transitions from frame _n_ − 6 to frame _n_ + 5. The 12 frames show the corners in each frame (from _n_ − 6 to _n_ + 5), and the candidate shot boundary is in frame _n_. In _n_ : 11759, _n_ denotes the frame number and 11759 is the total number of Harris corners in the _n_th frame, and the same for others.**

features such as pixel, edge, or histogram. Besides, it is expensive in computational complexity because most existing methods process all frames in a video.

However, frames that are far from the GT are very different, such as the first frame and the last frame in Figure 3.3. So if we skip sufficient frames each time, the GT can be regarded as CUT, which changes abruptly from one frame to a totally different one. Thus, we propose using a skip-interval method for finding shot boundaries accurately and quickly by reducing most of the processed frames.

3.3.2.1 Frame Similarity Measure with Mutual Information

Entropy is the measure of the amount of information that is missing before reception and is sometimes referred to as Shannon entropy. For random variable $X \in \{x_1, x_2, \ldots, x_n\}$, the entropy of X is defined as

$$H(x) = E(I(x)) \tag{3.2}$$

where e is the expectation function and $I(x)$ represents the amount of information of X. If p is the probability mass function of X, the entropy of X is rewritten as

$$H(X) = \sum_{i}^{n} p(x_i) I(x_i) = -\sum_{i=1}^{n} p(x_i) \log_b p(x_i) \tag{3.3}$$

while $p_i = 0$. We set $0 \log_b 0 = 0$, which is consistent with limitation $\lim_{\to 0^+} p \log p = 0$.

Meanwhile, the JE (Joint Entropy) of random variable X, Y is defined as

$$H(X, Y) = -\sum p_{XY}(x, y) \log p_{XY}(x, y) \tag{3.4}$$

where $p(x_i, y_j)$ is the joint probability density function.

According to knowledge of information theory, the mutual information of X and Y is defined as

$$I(X, Y) = -\sum_{i, j \in N, M} p(x_i, y_j) \log \frac{p(x_i, y_j)}{p(x_i) p(y_j)} \tag{3.5}$$

Besides, $H(X|Y)$ is also defined:

$$H(X|Y) = \sum_{y \in Y} p_Y(y) H(X|Y = 1) = -\sum_{x, y \in X, Y} p_{xy}(x, y) \log p_{xy}(y|x) \tag{3.6}$$

where $p_{XY}(y|x)$ denotes the conditional probability. The conditional entropy $H(X|Y)$ is the uncertainty in X given knowledge of Y.

Some important properties of the MI are

1. $I(X, Y) \geq 0$.
2. For both independent and zero entropy sources X and Y, $I(X, Y) = 0$.
3. $I(x, y) = I(Y, X)$.
4. The relation between the MI and the JE of random variables X and Y is given by $I(X, Y) = H(X) + H(Y) - H(X, Y)$, where $H(X)$ and $H(Y)$ are the marginal entropies of X and Y.
5. $I(X, Y) = H(X) + H(Y) - H(X, Y) = H(X) - H(X|Y)$
   $= H(Y) - H(Y|X)$.

In terms of the above-mentioned characteristic of mutual information, MI is not only the measurement metrics of two random variables' correlations, but also the information that X and Y share: it measures how much knowing one of these variables reduces uncertainty about the other. For example, if X and Y are independent, then knowing X does not give any information about Y and vice versa, so their mutual information is zero. At the other extreme, if X is a deterministic function of Y and Y is a deterministic function of X, then all information conveyed by X is shared with Y: knowing X determines the value of Y and vice versa. As a result, in this case the mutual information is the same as the uncertainty contained in Y (or X) alone, namely, the entropy of Y (or X). Moreover, this mutual information is the same as the entropy of X and the entropy of Y. (A very special case of this is when X and Y are the same random variable) (Wikipedia) https://en.wikipedia.org/wiki/mutualinformation.

Therefore, we adopt an idea similar to that in [191, 230] to define the frame difference with entropy and mutual information. In a video, we assume that the distribution of a pixel value is independent, and $p_X(x)$ means the probability of the gray value $x \in \{0, 1, \dots, 255\}$ in frame X. Thus, according to the entropy definition and joint entropy, $p_{XY}(x, y)$ in Equation 3.4 means the probability of a pixel pair with a gray value x in frame X and a gray value y in the corresponding pixel position in frame Y. Meanwhile, the mutual information of two frames X and Y can be obtained with property 5.

Let $V = \{F_1, F_2, \dots, F_N\}$ denote the frames of a video clip V. For two frames (F_x and F_y), we first compute their own entropies (H_x, H_y) and their joint entropy ($H_{x,y}$). The MI between them is given by Equation 3.5. If $I_{x,y}^R$, $I_{x,y}^G$, $I_{x,y}^B$ respectively represent the MI of each RGB component, we set $I_{x,y} = I_{x,y}^R + I_{x,y}^G + I_{x,y}^B$ as the MI between frames F_x and F_y.

3.4 Temporal Redundant Frame Reduction

Generally speaking, the human visual reaction is about 2 s, supposing the video frame rate is about 20–25 frames per second (fps); then a shot that can cause a visual reaction needs to last for 40–50 frames at least. Thereby, in order to efficiently reduce the processed frames and also not drop any boundaries between two shots, we assign the skip interval $s = 40$ frames in our approach. For a given video, F_x corresponds to the xth frames, and we calculate the similarity rate D between F_i and F_{i+s} as

$$D_{i,i+s} = \frac{\min(I_{i,i+1}, I_{i,i+s})}{\max(I_{i,i+1}, I_{i,i+s})} \tag{3.7}$$

where $I_{i,i+1}$ means the mutual information between F_i and F_{i+1}, and the same meaning of $I_{i,i+s}$ [230]. Generally, adjacent frames (i.e., F_i and F_{i+1}) in the frame sequence are nearly the same, so if $I_{i,i+1}$ is close to $I_{i,i+s}$, it means that F_{i+s} is similar to F_{i+1}, and consequently similar to F_i. Actually, if $D_{i,i+s} > 0.8$, we can assert that $I_{i,i+1}$ is close to $I_{i,i+s}$ and F_i is similar to F_{i+s}, and then we skip to process the next s frames. Otherwise, it means that there is a shot boundary existing between F_i and F_{i+s}. Thereby, we use a binary search to find the refine boundary in this range. First, we compute $D_{i,i+s/2}$ and $D_{i+s/2,i+s}$. If $D_{i+s/2,i+s} > 0.8$, it means that the boundary exists in the first half of F_i to F_{i+s}, and otherwise in the second half. Then, the analogical process continues in the first or second half of F_i to F_{i+s} to refine the boundary position until half of the range is only one frame.

After that, we have accelerated the process in both the spatial and temporal domains and nearly get all boundaries. However, because camera or object motion also makes frames change obviously when we skip the processing of several frames, the above process has also generated several false shot boundaries caused by camera motion, camera zoom in or out, and object motion. This is called motion-caused false shot boundary (MCFB). Actually, most false shot boundaries in our approach are MCFB, so in the next section, we introduce how to detect and remove these false boundaries.

3.5 Corner Distribution-Based MCFB Removal

After the above processing, we can get two types of candidate shot boundaries: the true shot boundary, CUT and GT, and the false alarm, MCFB. In this section we elaborate on how to use the corner distribution in frames to remove MCFBs.

Corner distribution means the scattered and changed status of a detected corner in frames, including the number of corners, their distributed positions, and the variety discipline. Because corners are the distinguishing feature of a frame, different frames have very different corner distributions. Through comparative analysis, we found that different candidate shot boundaries, such as CUTs, GTs, and MCFBs, always have different corner distributions, and this property can be used for removing MCFBs. For a GT boundary, the corner distribution of frames in a boundary is not similar to either frames in a forward shot or frames in a backward shot, but it is fused by both as shown in Figure 3.4. Because frames in a GT boundary change slowly and gradually, frames in or near the boundary are always overlaps of the forward and backward shots' content. Meanwhile, a GT boundary is an editing effect, the

Figure 3.4 Example of corner distribution in a CUT between frames 3722 and 3723.

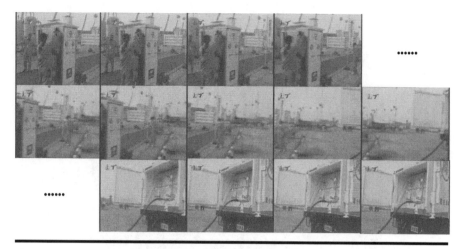

Figure 3.5   A sequence of frames in a camera motion.

details of which cannot be observed precisely by an audience. Associating with the aforementioned human visual reaction time, the GT duration is usually small, that is, 5–20 frames, and far less the visual reaction of 50 frames. For a CUT boundary, the frame change from one scene to another directly, the corner distribution is very abrupt and dramatic.

Finally, for MCFB, the audience can and should feel these camera or object motions. Thus, frames in MCFBs last for nearly hundreds of frames, longer than those in CUTs or GTs. For example, the camera motion lasts about 150 frames in Figure 3.5, but only 9 frames in the GT of Figure 3.3. A long-duration change also means a high similarity of adjacent frames. In Figures 3.3 through 3.5, we found that corner distribution changes obviously in a real shot boundary, but slowly and smoothly in MFCB. For example, in the GT boundary shown in Figure 3.3, there are 11,759 corners in F_n, but 6,113 and 5,957 corners in $F(n-5)$ and $F(n+5)$, respectively.

Therefore, for a candidate shot boundary position F_n, we compare its corner distribution with that in $F(n-c)$ and $F(n+c)$, where c is the interval (which is not bigger than 15 in our approach). If the corner distribution of F_n is similar to that of $F(n-c)$ and $F(n+c)$, the corner distribution is stable and consistent for frames near this candidate shot boundary, and we can assert the candidate shot boundary located in F_n is a MCFB. In the next paragraph, we will describe in detail how to judge if two frames are similar or not by corner distribution.

In order to judge if two frames are similar, it is convenient to compare the total Harris corner number of these frames. But it is inaccurate and crude to use only this

Block

Figure 3.6   Blocks in a frame of size $k \times l$.

simple feature, and we thus introduce local corner distribution. For two frames of size $K \times L$, local corner distribution is used to measure their similarity. The general steps are as follows:

1. Each frame is divided into $K/k \times L/l$ blocks, and each block is of size $k \times l$, as shown in Figure 3.6.
2. Corners in each block are detected by the Harris corner detector [47]. $Num_A(u, v)$ and $Num_B(u, v)$ denote the number of corners detected in block (u, v) of frames A and B, respectively (suppose $Num_A(u, v) < Num_B(u, v)$; otherwise, exchange the labels of A and B).
3. Then, we compare the corner distribution of blocks in the same position of the two frames. Each block position (i.e., (u, v)) generates a compare label $CR_{u,v}$ that represents the matching between the two blocks in this position, and the matching rules are as follows:

$$
CR_{u,v} = \begin{cases} 1 \text{ if } Num_A(u, v)/Num_B(u, v) > T_\theta \\ \\ 0 \text{ otherwise} \end{cases} \tag{3.8}
$$

When $CR_{u,v}$ equals 1, it corresponds to a match. T_θ is the threshold, which is empirically fixed with 0.82 in our approach. Finally, we count the proportion of blocks with a compare label $CR = 1$.

4. The proportion of the matching blocks is calculated as

$$MP = \frac{\sum_{u,v} CR_{u,v}}{k \times l} \qquad (3.9)$$

If the proportion MP is bigger than 85%, we assert that the two frames' local corner distributions are consistent and similar.

3.6 Experimental Results

To assess the performance of our approach, we compare the efficiencies in several aspects. Our test video set consists of four clips that represent different categories, as shown in Table 3.1.

We have done experiments on both the processing efficiency and accuracy of our approach, and an example of the segmented shots is shown in Figure 3.7. We first compare the efficiency of our approach by using a focus region with the method proposed in [230] by using all pixels. The result is shown in Table 3.2. Then we provide the comparative experimental results of intervals of 40 frames, 5 frames [45], and 20 frames based on histogram [111] and also the frame-by-frame method based on mutual information [230]. The performance of removing the MCFB by corner distribution is also proposed. The results are shown in Table 3.3.

In our approach, we segment each frame into a size of 15×15 to get the FR, and the skip interval is 40 frames. From Table 3.2, we can see that our method based on FR not only accelerates the process in the spatial domain and economizes time, but also has satisfactory accuracy. Too big of a value for the step interval will lose shot boundaries, while a small one does not really accelerate the SBD. Table 3.3 shows that almost all boundaries are detected by our approach with an interval of 40 frames and removing the MCFB, no matter if it is CUT or GT. It especially increases the accuracy for detecting GT. Compared with methods of frame-by-frame processing and methods of five-frame intervals, our approach really accelerates the process.

Table 3.1 Test Video Clips

Name	Duration (s)	Total Frames	Shot Number
Clip 1: News: CCTV news broadcast	3,324	83,100	692
Clip 2: Sports: Football	2,356	58,900	378
Clip 3: Movie: *Set Off*	1,802	41,446	511
Clip 4: Teleplay: 3 episode of *Marriage Battle*	2,072	51,800	497

Figure 3.7 Example of segmented shots in CCTV news.

Table 3.2 Test Video Clips and Comparisons of FR and Frame by Frame

		Clip 1	Clip 2	Clip 3	Clip 4
	FR-based precision	98.1%	93.5%	95.4%	95.1%
	FR-based recall	98.3%	93.4%	93.9%	97.1%
Time-consuming	Method of all pixels	808 s	2220 s	831 s	2040 s
	Method of focus region (15 × 15)	487 s	1332 s	510 s	1236 s

Table 3.3 Detection Accuracy and Duration Comparisons

Accuracy and Efficiency		Clip 1 Rec. Pre. (%)	Cost Time	Clip 2 Rec. Pre. (%)	Cost Time	Clip 3 Rec. Pre. (%)	Cost Time	Clip 4 Rec. Pre. (%)	Cost Time
Mutual information	Frame by frame	96.7 / 99.7	808 s	94.7 / 98.0	2220 s	94.2 / 98.5	831 s	93.7 / 94.8	2040 s
	5-frame interval	97.0 / 98.1	247 s	94.9 / 98.3	511 s	95.7 / 96.5	257 s	96.0 / 97.9	507 s
	40-frame interval	85.8 / 89.2	58 s	80.9 / 88.3	72 s	85.2 / 86.1	63 s	80.5 / 87.5	70 s
	FR and 40-frame interval and MCFB removing	97.1 / 99.3	54 s	94.9 / 98.3	76 s	95.7 / 96.5	69 s	96.0 / 97.9	84 s
Color histogram	20-frame interval	85.1 / 90.0	98 s	83.3 / 89.5	139 s	80.7 / 88.2	86 s	90.1 / 91.9	121 s

3.7 Summary

The high cost of most SBD techniques becomes an obstacle in video processing procedures, especially in real-time scenarios. In this chapter, we addressed this problem by decreasing both examined frames and pixels. First, only pixels falling into the FR regions were used for computing the frame difference. Second, we skipped several frames to find the boundary by a step-skip method, while most redundant frames were excluded for processing. By skipping several frames, we further detected more GTs, which is very difficult for SBD. Finally, while most false boundaries (i.e., MCFB) are introduced by the above process, the corner distribution of frames was used to remove them, and the experimental results show that our method not only sped up the SBD to save computational effort, but also archived satisfactory accuracy.

Chapter 4

Key Frame Extraction

The essential task in video analysis and indexing is to present an abstract of the entire video within a short clip, namely, to provide a compact video representation while preserving the essential activities of the source video. Nevertheless, most video cataloguing approaches are based on the selection of key frames within the shots of a video. Thus, key frame extraction is a very crucial step in video processing and indexing. Generally, a key frame is defined as one or several frames that typically represent the whole content in a small video clip, that is, a shot or a scene. In this chapter, we introduce several typical key frame extraction methods and a novelty key frame extraction method for feature movies.

4.1 Introduction

Recently, with the rapid advances in digital technologies, there has been a drastic increase in the creation and storage of video data on the Internet. Key frame extraction is derived from the nature of videos, as usually every video contains a lot of redundant information that can be removed to make the data more compact and valuable. Actually, key frame extraction is the static way to realize a video summary, and key frames in a video represent the most important and typical content of the video. The most common or direct key frame can be seen as the cover of a movie on the online video website. In all, key frame extraction is always used as a preprocessing step for most video processing or analysis applications, which suffer from the problem of processing a large number of video frames, especially in this Big Data century.

For key frame extraction, the challenge is how to find an optimal and minimal set of representative frames that converge all significant content or the real semantics of a scene; namely, key frames provide the abstraction for a whole video. Also, the extraction of key frames greatly reduces the amount of data required in video indexing and browsing as well as data storage. That is, the user can browse a video by viewing only a few highlighted frames.

*The basic principle of key frame extraction is to make the extracted frames or content as **representative** and **minimal** as possible.* The principle of representative is to make sure the important or crucial information is not discarded or moved, and the principle of minimal is to ensure that most of the redundant information is removed and the key frame set is a high-efficiency compaction of the original video. Here, we use the definition in [182] of a video key frame:

$$\mathcal{R} = \mathcal{A}_{Keyframe}(V) = f_{r1}, f_{r2}, \ldots, f_{rk} \tag{4.1}$$

where $\mathcal{A}_{Keyframe}$ denotes the key frame extraction procedure.

There is one more thing to declare: When a whole video contains very complex content and many semantic topics, key frame extraction is always done in a short clip with a single topic or semantic information shot generally.

Traditional key frame extraction techniques aimed to discard similar frames and preserve the frames that were different from one another and drawing attention to people. Thus, visual features are usually preferred as the principle for comparing the relational degree between frames. Current key frame extraction can be classified according to its various measurements of visual content complexity of a video shot or sequence. In the early years, the easiest key frame extraction method was to directly choose the first, middle, or ending frames as the key frames. The advantage of this was that it was simple and efficient with low computation complexity. Of course, the disadvantage is also obvious: The extracted key frames had a low correlation in visual content.

4.2 Size of the Key Frame Set

With the classification in [182], there are three types of mechanisms to fix the size of the key frame set: fixed as a known prior, left as an unknown posterior, and determined internally within the video processing applications. Nevertheless, there are also methods that use the combination of two types of mechanisms for the size of key frame set [49, 121]. For example, the procedure will stop when the number of key frames reaches a prior value or when a certain condition is satisfied (i.e., a posteriori).

Fixed as a known prior. In this mechanism, the size of the key frame set is given a constraint value before the experiment or procedure; it can also be named rate constraint key frame extraction. While a prior value of the size of a key frame set is k and the output of the key frame set is $\mathcal{R} = \{f(i_1), f(i_2), \ldots, f(i_k)\}$, the formula definition of fixed as a known prior set is shown as follows:

$$\{r_1, r_2, \ldots, r_k\} = \arg\min_{r_i}\{\mathcal{D}(\mathcal{R}, \mathbf{V}, \rho)|1 \leq r_i \leq n\} \tag{4.2}$$

where n is the number of frames in original video V and ρ is the extraction perspective. As the description in [182], ρ always refers to visual coverage, but it can also be the number of objects, number of faces, and so on. Therefore, with fixed as a known prior, the approach is usually suitable to extract key frames in a video without dramatic content variations; that is, a shot is captured with or without slight camera moving or tilt.

Left as an unknown posterior. If we do not know the number of extracted key frames until the process finishes in a specific application, the size of the key frame set can always be seen as this type. This type of set size is suitable for video sequences with plenty of visual variations or a lot of action and movements. Even so, for those video sequences with very highly dynamic scenes, a massive large number of key frames will be generated and result in inconvenience for the following video processing. Meanwhile, with too large of a number of key frames, it does not conform to the basic principle of minimal. With a specific dissimilarity tolerance, the formulation of the posterior-based key frame set size is defined as

$$\{r_1, r_2, \dots, r_k\} = \arg \min_{r_i} \{\mathbf{k} | \mathcal{D}(\mathcal{R}, \mathbf{V}, \rho) < \varepsilon, 1 \le r_i \le n\} \qquad (4.3)$$

where n, ρ, and $\mathcal{D}$ are with the same meaning as in the prior type.

Determined internally. This is a particular type of posterior approach because the appropriate number of key frames is determined before the extraction processing, especially for those clustering algorithm-based approaches [56, 58, 72].

4.3 Categories of Key Frame Extraction Methods

In this section, we detail the basic classification of those commonly used key frame extraction methods, according to the articles proposed by Xu et al. [199]. In Xu et al.'s opinion, the commonly used key frame extraction methods can usually be loosely classified into the following categories:

Sequential algorithms. In this category, a new key frame is determined when the difference between the new coming frames and the existing key frames exceeds the prior fixed threshold [92, 152]. For example, in [152], they detect multiple frames based on the visual variations in shots. Let $S_{\ddagger}$ refer to a shot, and $K_{\ddagger}$ means the set of key frames. First, the key frame set $K_{\ddagger}$ is initialized into empty. Then, the middle frame of $S_{\ddagger}$ is chosen to add to $K_{\ddagger}$ as the first key frame. After that, each frame in $S_{\ddagger}$ is compared to frames in $K_{\ddagger}$ with $M : N$ comparing. If the difference between the frame in $S_{\ddagger}$ and $K_{\ddagger}$ is bigger than the fixed threshold, then the frame in $S_{\ddagger}$ is added into $K_{\ddagger}$ as a new key frame.

Step 1: Select the middle frame to initialize $K_{\ddagger}$,

$$K_{\ddagger} \leftarrow \{f^{\lfloor \frac{a+b}{2} \rfloor}\} \qquad (4.4)$$

Step 2: For $r = a$ to b, if $\max(ColSim(r, k) < T, \forall f^k \in K_{\ddagger})$, then $K_{\ddagger}) \leftarrow \bigcup\{f^r\}$, where T is the similarity threshold that declares two frames to be similar or not. *ColSim* is the color similarity between two frames and is referred to as the histogram intersection distance. However, the sequential algorithms have the problem that the extracted key frames do not cover the video content acceptably.

Clustering-based algorithms. These algorithms look at the video frames as data points in the feature space at first, and then define the point-to-point distance. With the distance measurement, data points are clustered and key frames are chosen as those nearest to the cluster centers. An example for this type of algorithm is shown in. In [30], key frame extraction is performed by clustering frames of a shot into groups using an improved spectral clustering algorithm. Finally, the medoid of each group or cluster, namely, the frame of a group whose average similarity to all other frames is maximal, is characterized as a key frame. As described in [30], the spectral clustering algorithm are shown as below.

Suppose $H = H_1, H_2, \ldots, H_N$ is a set of frame feature vectors (color histogram) corresponding to the frames of a video, and the destination is to partition them into K groups.

1. Compute the similarity matrix $A \in \mathbb{R}^{NN}$ for the pairs of feature vectors of the dataset S, where each element of A is computed as follows:

$$\alpha(i, j) = 1 - \frac{1}{\sqrt{2}}\sqrt{\sum_{h \in bins} (H_i(h) - H_j(h))^2} \tag{4.5}$$

2. Define D to be the diagonal matrix whose (i, i) element is the sum of the elements of A's ith row and construct the Laplacian matrix $L = I - D^{-1/2}AD^{-1/2}$.
3. Compute the K principal eigenvectors $x_1, x_2, \ldots, x_K$ of matrix L to build an $N \times K$ matrix $X = [x_1, x_2, \ldots, x_K]$.
4. Renormalize each row of X to have a unit length and form matrix Y so that

$$y_{ij} = \frac{x_{ij}}{(\sum_j x_{ij}^2)^{\frac{1}{2}}} \tag{4.6}$$

5. Cluster the rows of Y into K groups using the Global k-means algorithm.
6. Finally, assign feature vector H_i to cluster j if and only if row i of matrix Y has been assigned to cluster j.

One of the shortcomings of these clustering-based methods is the high computational cost; another one is that the key frames obtained lose the temporal information of the original video. However, this kind of information is clearly helpful for quickly grasping the video content.

Shot-based algorithms. For this category, all processing is done in a shot. The shot is first split into finer segments (subshots) that are more homogeneous progressively, and then the key frames are extracted from these subshots. This type of algorithm is easier and more efficient since just a small number of frames are operated at a time. A typical example of this category is the novel key frame extraction approach proposed by Xu et al. [199]. They first segment the shot into subshots in terms of the content changes, and then an extraction representative frame in each subshot to form the key frame set.

More specifically, the Jensen divergence (JD) is used to define the difference of frames at first. With equal weights for probability distributions, the simple case of JD [214] is defined as

$$J D(p_1, p_2, \ldots, p_n) = F\left(\frac{1}{n}\sum_{i=1}^{n} p_i\right) - \frac{1}{n}\sum_{i=1}^{n} F(p_i) \tag{4.7}$$

where F is a concave function. After that, the JD values of two frames f_{i-1} and f_i are,

$$J D(f_{i-1}, f_i) = F\left(\frac{p_{f_{i-1}} + p_{f_i}}{2}\right) - \frac{F(p_{f_{i-1}}) - F(p_{f_i})}{2} \tag{4.8}$$

where $p_{f_{i-1}}$ and p_{f_i} are the respective probability distributions of f_{i-1} and f_i, which correspond to the normalized histogram distributions of the two frames. Meanwhile, an average of JDs with respect to all the members in a window of f_i is named the window-sized JD ($J D_w$):

$$J D_w(f_{i-1}, f_i) = \frac{1}{n_w}\sum_{j=i-\lfloor\frac{n_w}{2}\rfloor}^{j=i+\lfloor\frac{n_w}{2}\rfloor} J D(f_i, f_{i-1}) \tag{4.9}$$

where n_w is the size of the slide window.

Based on the above-defined variable, the gradients of JD values are used to partition a shot into several subshots.

$$\triangle_{J D}(f_i) = J D_w(f_i, f_{i+1}) - J D_w(f_{i-1}, f_i) \tag{4.10}$$

Meanwhile,

$$\triangle_{J D_w}(f_i) = \frac{1}{n_w}\sum_{j=i-\lfloor\frac{n_w}{2}\rfloor}^{j=i+\lfloor\frac{n_w}{2}\rfloor} \triangle_J D(f_i) \tag{4.11}$$

For each shot, the window-sized $J D$ gradient data $\triangle(J D_w)$ can be used for the subshot identification. Specifically, if $\|\triangle_{J D_w}(f_s)\| \geq \triangle_{J D_w}^*$, it means that there is an obvious content change around frame $f_s s$, while $\Delta^*_{JD_w}$ is the prefixed threshold. Starting from this "summit" $f_s s$, its first left and right frames, f_l and f_r, within this shot, where both $\|\Delta_{JD_w}(f_l)\|$ and $\|\Delta_{JD_w}(f_r)\|$ are close to zero, are found. Actually, the indexes l and r are obtained by

$$r = \max\{j \, \|\Delta_{JD_w}(f_i)\| < T^*_{JD_w} \bigwedge j < ss\},$$

$$r = \min\{j \, \|\Delta_{JD_w}(f_i)\| < T^*_{JD_w} \bigwedge j > ss\}, \tag{4.12}$$

where j is inside the considered shot and $T^*_{JD_w}$ is a threshold. Until now, the video frame sequence $[f_l, f_r]$ has referred to a subshot with an obvious content change, and the key frame has extracted in this subshot. For a subshot with the obvious changes, the key frame is selected to minimize the summed JD values between it and all the others; otherwise, the center frame of the subshot is selected as the key frame. Finally, with all those extracted key frames from the subshot, we get the key frame set and finish the key frame extraction.

Optimization-based algorithms. In addition, optimization-based algorithms also use feature space and its data points for the frames of a video. The data points optimizing a specifically designed objective function are selected as the key frames. The main drawback is that these algorithms are very time-consuming, conflicting with the goal of our book.

4.4 Key Frame Extraction Using a Panoramic Frame

In order to get the most representative features as well as to reduce the noise involved in redundant frames, we propose a novel key frame extraction method using a panorama frame. Recently, several key frame selection methods were proposed to choose an appropriate number of key frames, especially for a dynamic shot with a larger motion of actor or camera [152, 214]. However, these methods may satisfy the requirement of applications such as video summary, rather than the specific application of movie scene recognition in Chapter 7. That is because to compare with these applications, we need more comprehensive and representative features of a video scene (VSC) to distinguish different scene categories. While the panoramic frame obtained by video frame registration contains more completed features, Xiao et al. [193] introduced the problem of scene viewpoint recognition with panoramic images organized into 26 place categories. Thus, in this chapter, we also use the panoramic frame to construct our key frames.

The panoramic frame can be obtained by video registering. In this chapter, in order to obtain a more robust panoramic frame, we adopt the robust video registration (RVR) method proposed by Ghanem et al. [62] As described in [62], we get the panoramic frame by calculating the homography matrix between two frames, since with the homography matrix, we can map all the pixels in one frame to another to generate the panoramic frame.

Figure 4.1 Examples of key frames using panoramic frames.

More specifically, $F_t \in \mathbb{R}^{M \times N}$ means the frame at time t and h_t refers to the homography matrix from F_t to $\mathbf{F}_{t+1}$. We spatially transfer $\mathbf{F}_t$ to $\mathbf{F}_{t+1}$ using the operation of $\tilde{\mathbf{F}}_{t+1} = \mathbf{F}_t \circ \mathbf{h}_t$. Ideally, $\tilde{F}_{t+1}$ should be approximated to $\mathbf{F}_{t+1}$, and the error arising from outliers pixels, denoted as $\mathbf{e}_t = \tilde{\mathbf{F}}_{t+1} - \mathbf{F}_{t+1}$, should be assumed to be sufficiently sparse. So far, the video registration problem is transferred to estimating the optimal sequence of homography matrices that map consecutive frames and render the sparsest error (minimum ℓ_0 norm). Since this problem is NP-hard in general and nonconvex, especially due to the nonlinear constraints, the cost function is replaced with its convex envelope (ℓ_1 norm) as follows:

$$min_{\mathbf{e}_{t+1}} \|\mathbf{e}_{t+1}\|_1$$

$$s.t. : \mathbf{F}_t \circ \mathbf{h}_t = \mathbf{F}_{t+1} + \mathbf{e}_{t+1} \tag{4.13}$$

Although the objective function is convex, the equality constraint is still not convex. Thus, with a iteratively solved linearized convex problem, we begin with an estimation of each homography, denoted as $h_t^{(k)}$ at the $(k+1)$th iteration. In order to linearize the constraint around a current estimate of the homography, the current estimation will be $\mathbf{h}_t^{(k+1)} = \mathbf{h}_t^{(k)} + \triangle \mathbf{h}_t$. Thus, Equation 4.13 is relaxed to

$$min_{\triangle \mathbf{h}_t, \mathbf{e}_{t+1}} \|\mathbf{e}_{t+1}\|_1$$

$$s.t. : \mathbf{J}_t^{(k)} \triangle \mathbf{h}_t - \mathbf{e}_{t+1} = \delta_{t+1}^{(k)} \tag{4.14}$$

where $\delta_{t+1}^k = \mathbf{F}_{t+1} - \mathbf{F}_t \circ \mathbf{h}_t^{(k)}$ represents the error incurred at iteration k and $\mathbf{J}_t^{(k)} \in \mathbb{R}^{MN \times 8}$ is the Jacobian of $\mathbf{F}_t \circ \mathbf{h}_t$. Finally, the homography matrix is successfully extracted since the problem in Equation 4.14 becomes a linear problem that can be solved in polynomial time.

After that, the panoramic frame in each shot can be obtained with this estimated homography matrix, as can be seen in Figure 4.1. More details of the video registration and stitching can be seen in [62].

4.5 Summary

Key frame extraction is an essential process in video analysis and management, providing a suitable video summarization for video indexing, browsing, and retrieval. The use of key frames reduces the amount of data required in video indexing and provides a compact video representation while preserving the essential activities of the source video. Generally, a key frame is defined as one or several frames that can typically represent a shot or a scene. In this chapter, we introduced several typical key frame extraction methods and then proposed a novel panoramic frame-based key frame extraction method that is very suitable for video cataloguing, especially in feature movies.

Chapter 5

Multimodality Movie Scene Detection

With rapid advances in digital technologies, feature movies are a large portion in the powerful growth of videos. Although, with shot boundary detection, we access the logical unit or basic unit for a video, we still cannot index or browse a video (especially a movie) in the way of human understanding. Fortunately, in order to feasibly browse and index these movies, movie scene detection is another important and critical step for video cataloguing, video indexing, and retrieval.

5.1 Introduction

A scene is a video clip comprised of a cluster of shots with the same similarity or presenting the same topic. Moreover, these shots have a strong spatial and temporal relation, and there is also a high relation of low-level features (such as color, edge, and texture) and semantic features. A shot is a bunch of frames captured by the camera at one time; thus, generally, shot boundary detection is usually implemented by physical segmentation based on low-level features. Yet, scene detection, or scene segmentation, mainly focuses on semantic relations; namely, it is a type of semantic segmentation, and thus there will be more challenges. More generally, a scene is also called a video clip, video plot, or story unit. There are three types of definitions for scene:

1. Scene in the same time and same place: This is a video clip composed with overlapped shots captured in the same place and same time, for example, dialogue scenes, fighting scenes, and so on.

2. Scene in the same time and continuous place: This refers to video clips composed of continuous shots captured in a continuous place and time duration, for example, a chase scene.
3. Scene with the same topic: This is an aggregate of a series of shots with the same semantic topic, for example, news scenes and documentary scenes.

Here, we mainly discuss the first two scenes, which are also frequently seen in movies and TV series.

A video or movie scene is the basic unit for video production and editing. Meanwhile, the change of scene scenario is also the crucial aspect for the development of a story plot. Obviously, shot boundary detection can help to organize and manage the video content, but shot as a video unit without sufficient narrative and semantic features. Therefore, if a shot is looked at as the physical structure unit of a video, a scene can be seen as the semantic structure unit, which can reflect the semantic hierarchy and has a more abstract and summary ability. In addition, a video scene is constructed on a video shot with consideration of the semantic relation. Video scene detection not only is useful for semantic content analysis and video content concentration, but also can help in video indexing, retrieval, and browsing.

Scene detection is the fundamental step for the efficient accessing and browsing of videos. In this chapter, we propose segmenting movies into scenes that utilize fused visual and audio features. The movie is first segmented into shots by an accelerating algorithm, and the key frames are extracted later. While feature movies are often filmed in open and dynamic environments using moving cameras and have continuously changing contents, we focus on the association extraction of visual and audio features. Then, based on kernel canonical correlation analysis (KCCA), all these features are fused for scene detection. Finally, spatial–temporal coherent shots construct the similarity graph, which is partitioned to generate the scene boundaries. We conducted extensive experiments on several movies, and the results show that our approach can efficiently detect the scene boundaries with satisfactory performance.

5.2 Related Work

Generally, scenes are defined as sequences of related shots chosen according to certain semantic rules. Shots belonging to one scene are often taken with a fixed physical setting; the continuity of ongoing actions performed by the actors is also seen as a scene.

Over the last decades, many scene detection methods have been proposed. Tavanapong and Zhou [180] introduced a stricter scene definition for narrative films and visual features from selected local regions of extracted key frames. Then features were compared by the continuity editing techniques for filmmaking. Chasanis et al. [30] clustered shots into groups, and then a sequence alignment algorithm was applied to detect scenes when the pattern of shot labels changed. In [154], a weighted undirected graph called shot similarity graph (SSG) was constructed, and then scene detection was transformed into a graph partitioning problem.

However, compared to news and sports videos, scene detection for movies or teleplays is more challenging because of the complex backgrounds and variations of content. Meanwhile, movies and teleplays are different from the formal two types of videos with plenty of domain knowledge. Generally, with the different types of features used, we classify existing scene detection methods into the following classes:

1. **Merging-based scene detection**. This group looks at scene detection as a merge process from bottom to top, namely, to merge shots with visual similarity gradually. In [151], Rasheed and Shah proposed a method with two stages. In the first stage, the forward shot correlation is calculated with color similarity to get the candidate scene boundary. In order to deal with oversegmentation involved in the first stage, the oversegmented candidate scenes are merged together with shot length and motion features. In [30], the shot cluster is constructed with visual similarity and each shot cluster is assigned a tag. Then, the sequence alignment algorithm is adopted to obtain the scene boundary by defining and analyzing shot change patterns.

2. **Splitting-based scene detection**. This type of method is just the opposite of the formal one. That is, it looks at the whole video as one scene at first, and then splits this video into several small scenes from top down, gradually. In [208], a scene transfer graph is constructed with shot similarity, and the whole graph is looked at as a scene. Then, with the graph segmentation algorithm, the original graph is segmented into several subgraphs, and each subgraph corresponds to a scene.

3. **Model-based scene detection**. Except for the above-mentioned two groups, the statistic models are widely used to do scene detection. For example, Tan and Lu [177] assumed that shots in the same scene will have high similarity and set each scene as a Gaussian mixture model to do scene detection. However, parameter selection it is still an open question for the model-based methods.

However, most of the above-mentioned methods merely exploit visual information, mostly the color features. There were also several multimodality feature-based methods. Rasheed and Shah [153] incorporated the shot length and motion contents to analyze scene properties. Zhou and Zhu [210] analyzed both the auditory and visual sources to semantically identify video scenes. In [100], an enhanced set of eigen-audio frames was created and visual information was used to align audio scene change indications with neighboring video shot changes. In these mentioned multimodality scene detection methods, a nontrivial issue is how the audio information is integrated with the visual information. While KCCA is very useful to improve the performance of multimodality recognition systems that involve modalities with a mixture of correlated and uncorrelated components, we proposed integrating features

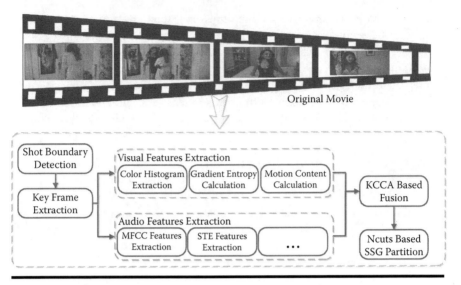

Figure 5.1 Flowchart of the proposed scene detection scheme.

of colors, motion contents, edges, and audio using the KCCA fusion mechanism for scene detection, as illustrated in Figure 5.1.

5.3 KCCA and Feature Fusion-Based Method

5.3.1 Shot Boundary Detection

First, an accelerating shot boundary detection method from our previous work [60] was adopted to segment movies into shots efficiently. Concretely, for each frame, the more the pixels near the center of the frame, the more important they are. Thus, the focus region (FR) in each frame is defined. Furthermore, using a skip interval of 40 frames not only speeds up the detection, but also finds more gradual transitions. Besides, the camera- and object-caused motions are detected as candidate shot boundaries, and using corner distribution analysis, all of them are excluded as false boundaries. These camera- or object-caused motions are also used in key frame extraction.

5.3.2 Key Frame Extraction

In order to choose an appropriate number of key frames, widely used middle frames may well represent a static shot with little actor or camera motion; however, a dynamic shot with higher actor or camera motion may not be represented adequately. So, we use a variant of [154] for key frame selection. In addition, to initialize the key frame set K, besides all the middle frames, each $5\times$ frame in the range of the detected motions is also contained. Actually, we use not only the color histogram, but also mutual information of frames to define the similarity between two frames. Mutual information-based similarity takes into account the mutual relations of corresponding pixels in two frames, as follows:

$$D_{i,i+s} = \frac{min(I_{i,i+1}, I_{i,i+s})}{max(I_{i,i+1}, I_{i,i+s})} \tag{5.1}$$

where $I_{i,i+1}$ means the mutual information between frames F_i and F_{i+1}, and the same meaning of $I_{i,i+s}$.

5.3.3 Audiovisual Feature Extraction

Descriptions in terms of hue/lightness/chroma or hue/lightness/saturation are often more relevant, so a 16-bin HSV normalized color histogram is computed for each frame with 8 bins for hue and 4 bins each for saturation and value. However, two different frames may have the same color histogram, so another visual feature, gradient entropy, is introduced.

First, for each key frame, a gradient magnitude $m(x, y)$ and orientation $\theta(x, y)$ in pixel (x, y) are calculated as

$$m(x, y) = \sqrt{(I(x + 1, y) - I(x - 1, y))^2 + (I(x, y + 1) - I(x, y - 1))^2} \tag{5.2}$$

$$\theta(x, y) = \tan 2\frac{I(x, y + 1) - I(x, y - 1)}{I(x + 1, y) - I(x - 1, y)} \tag{5.3}$$

where $I(x, y)$ is the gray value in pixel (x, y). After that, we can also get the gradient orientation histogram $\mathcal{O}$ based on $\theta(x, y)$. We map the whole orientation into 180 bins, and each bin's value is calculated by a weighted voting of the gradient magnitude. Then, we calculate the gradient entropy (GE) with consideration of the entropy theory. We first get the probability $\mathcal{O}(i)$ for orientation bin i as

$$\mathcal{O}(i) = \frac{o(i)}{\sum_{j=1}^{1} 80o(j)} \tag{5.4}$$

where $o(i)$ is the number of pixels in bin i. At last, we get the gradient entropy GE in terms of the orientation histogram $\mathcal{O}$ as

$$GE = -\sum_{i=1}^{180} \mathcal{O}(i) \log \mathcal{O}(i). \tag{5.5}$$

Actually, we get three gradient entropies: GE_r, GE_g, and GE_b in R, G, and B channels, respectively.

We have obtained 16 dimensions of the HSV histogram feature vector and 3 dimensions of the gradient entropy feature vector, in total 19 dimensions of visual features. Actually, in order to measure the visual feature of each shot, we also extracted the global motion description feature [153] and formed a 20-dimension feature vector for each key frame as the visual feature.

Generally, visual features can well represent the video content to some extent. Nevertheless, because of large-scale camera motion, a scene always concludes several different physical places, and single visual feature cannot do scene detection well in this situation. However, the audio content of a scene in the above-mentioned scenario is always consistent. Therefore, except for visual features, we also extracted and used the audio features.

When visual features can be extracted in the representative key frames, there are several audio clips in a video scene as well, and there is no exact corresponding relation between the audio and key frames. If the audio and visual features cannot correspond to each other, we cannot do feature fusion in the following processes. Thus, we first sample the corresponding audio into a 15 ms audio frame. Then, the forward and backward five audio frames with exact time-stamps corresponding to the key frame are extracted and grouped to form the audio feature.

In order to get the unitized audio feature vector, the audio feature extraction method proposed by Wang et al. [185] is adopted to generate a 43-dimension audio feature vector, including mel-frequency cepstral coefficients (MFCCs) and their delta and acceleration values (36-dimension feature), the mean and variance of the short-time energy (STE) log measure (2-dimension feature), the mean and variance of the short-time zero-crossing rate (ZCR) (2-dimension feature), the short-time fundamental frequency (or pitch) (1-dimension feature), the mean of spectrum flux (SF) (1-dimension feature), and the harmonic degree (HD) (1-dimension feature). Finally, the mean features of the 10 audio frames were calculated for the corresponding key frame. Until now, for each visual key frame, we have extracted 20-dimension visual features and 43-dimension audio features, and then we fused these features with KCCA.

5.3.4 KCCA-Based Feature Fusion

Different features extracted from the same pattern always represent characteristics of different aspects. With optimal combining or fusion, the unique description of each feature can remain, and the pattern can be more comprehensively described. Therefore, feature fusion is very useful and efficient for classification and recognition problems. As depicted in [232] there are three types of popular fusion schemes. The first one, centralized data fusion or information fusion, is to assimilate and integrate the information derived from multiple feature sets to directly get the final decision.

The second one is to make the first individual decision based on different feature sets, and then reconcile and combine them into a global decision, named distributed data fusion or decision fusion. The last one, feature fusion, is to produce a new fused feature set based on the given multiple feature sets, which is more robust for classification. In this chapter, we proposed a KCCA-based feature fusion approach to efficiently fuse visual and audio features.

Canonical correlation analysis (CCA) can be seen as solving the problem of finding basis vectors for two sets of variables such that the correlations between the projections of the variables onto these basis vectors are mutually maximized [77]. CCA is also adopted as a feature fusion method [136, 174]. Specifically, canonical correlation analysis is to map one object or pattern into two different feature spaces F^m and F^n to get two different feature sets X and Y, where X and Y are two random vectors with zero mean. According to the definition of CCA, the objective is to project X onto a direction ω_x as $X^* = \omega_x' X$ and Y onto a direction ω_y as $Y^* = \omega_y' Y$ and maximize the correlation between them as

$$\rho = \frac{E\,[X^*Y^*]}{\sqrt{E\,[(X^*)^2]\,E\,[(X^*)^2]}} = \frac{E\,[\omega_x' XY' \omega_y']}{\sqrt{E\,[\omega_x' XY' \omega_x]\,E\,[\omega_y' YY' \omega_y']}}$$

$$= \frac{\omega_x' C_{xy}\omega_y}{\sqrt{\omega_x' C_{xx}\omega_x \omega_x' C_{yy}\omega_y}} \tag{5.6}$$

where C_{xx} and C_{yy} are the covariance matrices of the trading data X and Y, C_{xy} is the cross-covariance matrix, and $E\,[\,]$ is the expectation value. To maximize ρ, the covariance matrix should satisfy the constraints as

$$\begin{cases} \omega_x' C_{xx}\omega_x = 1 \\ \\ \omega_y' C_{yy}\omega_y = 1 \end{cases} \tag{5.7}$$

For this optimization problem, the Lagrange multiplier is involved:

$$\max_{\omega_x,\omega_y} \omega_x' C_{xy}\omega_y - \frac{\lambda_x}{2}(\omega_x' C_{xx}\omega_x - 1) - \frac{\lambda_y}{2}(\omega_y' C_{yy}\omega_y - 1) \tag{5.8}$$

Therefore, to set the derivative of ω_x and ω_y to 0,

$$\begin{cases} C_{xy}\omega_y - \lambda_x C_{xx}\omega_x = 0 \\ \\ C_{yx}\omega_x - \lambda_y C_{yy}\omega_y = 0 \end{cases} \tag{5.9}$$

where C_{xy} and C_{yx} are the cross-covariance matrices of X and Y, and they are equal to each other. Multiply ω'_x and ω'_y in the left of the two equations in Equation 5.9 get the subtraction of them as

$$\lambda_x \omega'_x C_{xx} \omega_x - \lambda_y \omega'_y C_{yy} \omega_y = 0 \qquad (5.10)$$

Finally, Equation 5.7 is substituted into the above equation, and we can get $\lambda_x = \lambda_y = \lambda$. Since C_{xx} and C_{yy} are positive-definite matrices, we can further derive

$$\begin{cases} C_{xy} C_{yy}^{-1} C_{yx} \omega_x = \lambda^2 C_{xx} \omega_x \\ \\ C_{yx} C_{xx}^{-1} C_{xy} \omega_y = \lambda^2 C_{yy} \omega_y \end{cases} \qquad (5.11)$$

By solving the above equation, we can get each eigenvalue $\lambda_i > 0$ and its corresponding ω^i_x and ω^i_y, where $(i = 1, 2, \ldots, d, d = min(m, n))$ and d are the dimensions.

Nevertheless, there is the assumption of a linear relationship for different modality features, and this assumption is not suitable for many scenes. Therefore, we consider the kernel view of CCA, namely, kernel canonical correlation analysis (KCCA). KCCA is the powerful nonlinear extension of CCA to correlate the relation between two multidimensional variables, and it offers an alternative solution by first projecting the data into a higher-dimensional feature space,

$$K(x, y) = \langle \phi(x).\phi(y) \rangle \qquad (5.12)$$

and the Gaussian radial basis function,

$$K(x, y) = \exp\left(-\frac{\|x - y\|^2}{2\sigma^2} \right) \qquad (5.13)$$

is used, where σ is the width parameter to control the radial scope.

Now, when we get the correlation efficiencies ω_x and ω_y, we can fuse the extracted 20-dimension visual feature X and 43-dimension audio feature Y as

$$\mathcal{Z} = [\mathcal{X} \quad \mathcal{Y}] = \begin{pmatrix} w_x \\ w_y \end{pmatrix}^T \begin{pmatrix} X \\ Y \end{pmatrix} \qquad (5.14)$$

where $\mathcal{Z}$ is the fused feature, $w_x = (w^1_x, w^2_x, \ldots, w^d_x)^T$, $w_y = (w^1_y, w^2_y, \ldots, w^d_y)^T$. So far, we have a more comprehensive fused feature $\mathcal{Z}$.

5.3.5 Scene Detection Based on the Graph Cut

Graphic theory-based partition is a bottom-up partition method that is usually used for structure partition. There are many graph partition schemes, such as minimum cut, normalized cut (N-cut), and min-max cut. In these schemes, the N-cut has many more advantages, since it provides a standardized partition rule that will not result in the problem that the partition biases to the small subgraph. Graph partition techniques are widely used for scene detection with the construction of a graph in which each mode represents a shot, and the edges are weighted by their similarity or affinity [167]. First, the similarity between shots s_i and s_j is defined as

$$ShotSim(s_i, s_j) = \max_{p \in K_i, q \in K_j} Av\,Sim(p, q) \tag{5.15}$$

where p and q are the key frames in the key frame sets K_i and K_j, respectively. Meanwhile, $Av\,Sim(p,q)$ is the Euclidean distance of the fused audiovisual feature vectors in two key frames.

A weighted undirected graph $G = (V, E)$ is constructed with all shots, where the node set V denotes the shots and the weight of edge in set E denotes the node distance. Therefore, scene detection means seeking the optimal partition $V_1, V_2, \ldots, v_M$ of V, which maximizes the similarity among the nodes of each subgraph (V_i, E_i) and minimizes the cross-similarity between any two subgraphs. Namely, it partitions the node set V into subset V_i, where $\bigcup V_i = V$, and the intersection of any two subsets is a null set. Thus, with several bipartite segmentations of G, we can finally get all the subset, and each subset refers to a scene, which means we realize the scene detection.

In this chapter, the N-cut is employed to partition the shot similarity graph. The optimal disassociation is the minimum cut cost as a fraction of the total edge connections to all the nodes in the graph, called N-cuts.

$$Ncut\,V_1, V_2 = \frac{cut(V_1, V_2)}{assoc(V_1, V)} + \frac{cut(V_1, V_2)}{assoc(V_2, V)} \tag{5.16}$$

and

$$cut(V_1, V_2) = \sum_{v_1 \in V_1, v_2 \in V_2} w(v_1, v_2) \tag{5.17}$$

$$assoc(V_1, V) = \sum_{v_1 \in V_1, v \in V} w(v_1, v) \tag{5.18}$$

where w is the node distance between v_1 and v_2,

$$w(v_i, v_j) = exp\left(-\frac{(m_i - m_j)^2}{N^{\frac{1}{2}}\sigma^2}\right) \times ShotSim(v_i, v_j) \tag{5.19}$$

where σ is the standard deviation of shot duration in the entire video, m_i and m_j are the middle frame numbers of shots v_i and v_j, and N is the shot number. The details of the N-cut-based graph partition can be found in [154, 222].

Table 5.1 Summary of Test Dataset

Movies	Duration (s)	# Frames	# Shots	# Scenes
LFH	3, 324	83, 100	1, 308	35
CA	2, 353	58, 825	942	28
GLS	2, 835	70, 875	1, 126	30
PF	3, 715	92, 875	1, 611	41

5.4 Experiment and Results

To evaluate the performance of our approach, we experimented with four movies, including three Hollywood movies, *Love at First Hiccup* (LFH), *City of Angels* (CA), *The Goods: Live Hard, Sell Hard* (GLS), and a Chinese movie, *The Piano in a Factory* (PF), as can be seen in Table 5.1.

In order to evaluate different approaches with our approach, we used recall and precision measures as well as the *F*-measure with the combination of recall and precision:

$$FM = \frac{2 \times Recall \times Precision}{Recall + Precision} \tag{5.20}$$

The approach proposed by Rasheed and Shah [154] is a very classic scene detection method, as well as the comparison method used most in the literatures, but it only considers the visual feature. Kyperountas et al. [100] proposed combining the enhanced audio feature with the visual feature to do scene detection, and they achieved good performance. Therefore, we conducted an experimental comparison of our approach and the above-mentioned two methods considering the criteria of recall, precision, and *F*-measure; the results are shown in Table 5.2.

In the experiment, the four test movies were quickly and accurately segmented into shots and the key frames were also extracted. Then, the visual features of color, histogram, gradient entropy, and motion contents, as well as the 43-dimension audio features, were extracted for feature fusion. In order to fuse all these features using the KCCA algorithm, the Gaussian kernel in Equation 5.13 was used with parameter $\sigma = 1e6$.

In order to evaluate the results of our KCCA-based multimodality scene detection approach, Table 5.2 lists the comparison results of our approach and the other two methods. The method of [154] merely exploited the unimodality of visual features; thus the performance was not satisfactory based on the three criteria compared with the multimodality-based method. This is because when there are many object and camera motions, visual features cannot describe the consistency of a scene, resulting in false positives and false negatives. However, the audio consistency is more obvious and stable than the visual consistency; thus, with a combination of audio and visual

Table 5.2 Comparison of Proposed Approach with Rasheed and Shan's [154] and Kyperountas et al. [100] Methods

Movie Name	LFH	CA	GLS	PF	Total
Ground truth	35	28	30	41	134
Proposed Approach					
Detected	38	30	31	44	143
Recall (%)	83.0	85.7	83.3	87.8	**85.1**
Precision (%)	76.3	80.0	80.6	82.0	**80.0**
FM (%)	79.5	82.8	81.9	84.8	**82.5**
Method of Rasheed and Shah [154]					
Detected	45	39	33	53	170
Recall (%)	68.6	75.0	66.7	80.5	73.1
Precision (%)	53.3	54.0	60.6	62.3	57.6
FM (%)	60.0	62.8	63.5	70.2	**64.4**
Method of Kyperountas et al. [100]					
Detected	39	33	27	49	148
Recall (%)	80.0	82.1	66.7	80.5	77.6
Precision (%)	71.8	69.7	74.1	67.3	70.3
FM (%)	75.7	75.4	70.2	73.3	73.8

features, more satisfactory results were achieved. The method of [100] aligned audio scene change indications with neighboring video shot changes, while our approach took into account the visual features of color and gradient, motion contents, and several audio features for KCCA-based fusion. Although the method of [100] can deal with news video well, for movie content, the results of our approach were more robust and promising, as shown in Table 5.2.

5.5 Summary

It is still a challenging problem to robustly detect diverse movie scene changes. In this chapter, we addressed this problem using KCCA to fuse multimodality features in feature movies. First, we extracted the multimodality features, including the visual color histogram, gradient entropy, motion contents, and several audio features. Then, we used KCCA to fuse all of these extracted audiovisual features. Finally, the popularly used shot similarity graph partition method based on N-cuts was introduced to detect the scene boundary, and the experimental results show that our approach achieved satisfactory performance.

Chapter 6

Video Text Detection and Recognition

6.1 Introduction

Video text often contains plenty of semantic information that is significant and crucial for video retrieval and cataloguing. Generally, video text can be classified into superimposed text and scene text. The superimposed texts refer to texts added by the video editor in postediting, and these texts are the condensed semantic description of the video clips. The scene text is inherent text in the video captured by the video camera, and it is always an integral component of the background. Generally, scene text detection and recognition is very important in surveillance video analysis and the computer vision area, and superimposed text detection and recognition is always used for broadcast video analysis. While superimposed text is a sort of human-summarized semantic content with condensed meaning, it is perfect for video summary and cataloguing. Hence, in this book, we only pay attention to superimposed detection and recognition. Here, video text refers to superimposed text.

Video text detection and extraction are two crucial preprocessing steps for video text recognition. The text region is located with text detection. After that, because the superimposed text is always embedded in a complex background, text extraction is necessary to extract the text contour from the background. Finally, the real text recognition is done by an optical character recognition (OCR) system most of the time. However, text extraction is very crucial, and OCR-based recognition cannot get a satisfactory performance in the original text region with a complex background. For example, while we randomly selected 100 video text regions from news video, their recognition accuracy can be floated in the 60%–94% range by directly inputting to TH-OCR (typical Chinese character recognition software). That is, there is still a lot of room for improvement in recognition accuracy, and video text extraction is one of the ways for that.

In this chapter, we briefly introduce the ideas of video text detection, extraction, and recognition. Then, we depict a whole implementation process where, when a video is input to our system, we can output the texts with some existing methods. Actually, with this whole implementation process, we describe an interesting application used in basketball named highlight extraction with the text detection and recognition results (shown in Chapter 9).

6.2 Implementation of Video Text Recognition

In this section, we introduce a specific realization of video text recognition with some of our previous work and other existing methods. The whole implementation contains three processes: detection, integration extraction, and recognition. However, text recognition is always realized by some off-the-shelf OCR software; thus, we mainly discuss video text detection and extraction.

6.2.1 Video Text Detection

Superimposed texts are always horizontally and vertically aligned in videos. Moreover, different language characters have different appearances, which show different edge [159] and texture [3] characteristics in video. That is, the multilingual and multialignment text is pervasive in real video. Hence, video text detection is a difficult and challenging issue due to multiple language characters and multiple text alignments, as well as its being embedded in a complex background.

Despite that, many methods have been proposed for video text detection in the last decade [34, 61, 139], and some of them have achieved good performance. Most existing methods can be roughly classified into three categories: (1) text detection based on texture [134, 205], (2) text detection based on connected components (CCs) [93, 106], and (3) text detection based on color information [128, 146].

In texture-based methods, the embedded superimposed texts are looked at as particular texture patterns that are distinguished from the background in the video frame. Generally, the candidate text region is identified by either a supervised or unsupervised classifier. As examples of such methods, in [93], the intensities of the raw pixels that make up the textural pattern were input directly into the support vector machine (SVM) classifier, and text regions were detected by applying a continuously adaptive mean shift (CAMSHIFT) algorithm to the results of the texture analysis. The authors in [27] applied the Sobel edge detector in all Y, U, and V channels and then analyzed the invariant features, such as edge strength, edge density, and edge horizontal distribution. A statistical-based feature vector is proposed by Liu et al. [117] using the Sobel edge and K-means algorithm for text region detection.

In addition, there are also some methods using the frequency domain features. For example, in [63] and [106] the wavelet feature is adopted to detect aligned texts in an image, and also Zhong et al. [224] applied the discrete cosine transform to extract frequency features for text detection.

Meanwhile, another important category of methods is connected component-based text detection, which takes into consideration that text regions share similar properties, such as color and distinct geometric features, and have close spatial relationships. In [176], the candidate text regions are detected with a Canny edge detector, and then region pruning is carried out by means of an adjacency graph and some heuristic rules based on local component features. In [205], a multiscale wavelet energy feature was employed to locate all possible text pixels, and then a density-based region growing method was developed to connect these pixels into text lines. In [134], a stroke filter was used to detect all candidate stroke pixels, and then a fast region growing method was developed to connect these pixels into regions that were further separated into candidate text lines by a projection operation. In [52], the stroke width of each pixel was defined, and the stroke width transform (SWT) image was generated. After that, the SWT values were grouped with their ratios to produce the connected components.

In addition, there are still some methods that were not included in the above-mentioned two categories. For example, a hybrid approach is proposed in [145] where they first detected text regions in each layer of an image pyramid and then generated the candidate text components with a local binarization by projecting the text confidence and scale information back to the original image. Even so, most of the existing methods can be classified into these two categories, and we introduce the motion perception field (MPF) to locate the text region, especially with continuous video frames, as introduced in our previous work [85].

Video text always has a special motion pattern different from the background: the same text will keep up the location in a continuous video frame. Therefore, we propose a motion perception field to describe the special motion of video text and also detect text regions.

6.2.1.1 Motion Perception Field

We define the motion perception field to represent the motion pattern of video text. Superimposed video text always shows a distinct motion pattern; therefore, with motion pattern analysis, the video text can be better detected in a complex background. According to the analysis in [139], Lyu et al. conclude that superimposed texts have four language-independent characteristics: contrast, color, orientation, and stationary location. Although video text contains characters of different languages, the proposed motion perception field can perform well in a multilanguage environment. In addition, superimposed video always last for 1–2 s in the same location of a video frame. Hence, superimposed text is also named as nonmotion text (as shown in Figure 6.1a), whereas scene text is always labeled as dense motion

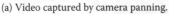

(a) Video captured by camera panning.   (b) Video captured by camera zooming.

(c) Video captured by a stationary camera.

Figure 6.1   Motion field on adjacent frames.

vectors (as shown in Figure 6.1b) and is called motion text. Besides, the scene text in Figure 6.1c is relatively stable because of the stability of the camera; it is called slight-motion text. We define the MPF as three types, DMPF, SMPF, and NMPF, which can represent the motion patterns of motion text, slight-motion text, and nonmotion text on consecutive frames, respectively.

Video text is a kind of special object, that is, rigid object in video. Thus, we obtain the motion vector field of adjacent frames by motion detection, and then the motion pattern on consecutive frames can be acquired through the fusion of multiframe motion vectors. In fact, the multiframe motion vector formed the motion perception field.

6.2.1.1.1   Motion Block Attribute on Adjacent Frames

The optical flow calculates the motion field between two frames that are taken at the times t and $t + \delta t$ at every pixel position. The optical flow method is based on local Taylor series approximations of the image signal. They use partial derivatives with respect to the spatial and temporal coordinates. Actually, a pixel at the location

(x, y, t) with the intensity $I(x, y, t)$ will have moved by δx, δy, and δt between the two frames, and the image constraint equation is shown as

$$I(x, y, t) = I(x + \delta x, y + \delta y, t + \delta t) \tag{6.1}$$

Given that the movement is very small, the image constraint at $I(x, y, t)$ with Taylor series can be developed to get

$$I(x + \delta x, y + \delta y, t + \delta t) = I(x, y, t) + \frac{\partial I}{\partial x}\delta x + \frac{\partial I}{\partial y}\delta y + \frac{\partial I}{\partial t}\delta t + h.o.t \tag{6.2}$$

A pixel at the location (x, y, t) with the intensity $I(x, y, t)$ will have moved by δx, δy, and δt between the two frames, and the following constraint equation can be given:

$$\frac{\partial I}{\partial x}\frac{\partial x}{\partial t} + \frac{\partial I}{\partial y}\frac{\partial y}{\partial t} = -\frac{\partial I}{\partial t} \tag{6.3}$$

where $V_x = \frac{\delta x}{\delta t}$ and $V_y = \frac{\delta y}{\delta t}$ are the x and y components of the velocity or optical flow of $I(x, y, t)$.

We use Singh's two-stage matching method [169] to get motion block velocity $V_b = (V_x, V_y)$. The first stage is based on the computation of the sum of squared difference (SSD) values with three adjacent band-pass-filtered images, I_{-1}, I_0, and I_{+1}.

$$SSD_0 x, d = SSD_{0,1}(x, d) + SSD_{0,-1}(x, -d) \tag{6.4}$$

where $SSD_{i,j}$ is given as

$$SSD_{i,j}(x, d) = \sum_{b=-n}^{n} \sum_{a=-n}^{n} W(a, b)[I_i(x + (a, b)) - I_j(x + d + (a, b))]^2$$

$$= W(x) \times [I_i(x) - I_j(x + d)]^2 \tag{6.5}$$

where W denotes a discrete two-dimensional (2D) window function and shift $d = (d_x, d_y)$ takes on integer values.

Adding two-frame SSD surfaces to form SSD_0 tends to average out spurious SSD minima due to noise or periodic texture. Then Singh converts SSD_0 into a probability distribution by

$$R_c(d) = e^{-m \times SSD_0} \tag{6.6}$$

where $m = -ln(0.95)/(min(SSD_0))$.

The motion block velocity $V_b = (V_x, V_y)$ is computed as the mean of this distribution (averaged over the integer displacements d).

$$V_x = \frac{\sum R_c(d)dx}{\sum R_c(d)} \tag{6.7}$$

$$V_y = \frac{\sum R_c(d)dy}{\sum R_c(d)} \tag{6.8}$$

Based on V_b, we define b as a motion block attribute on adjacent frames, which is computed as follows:

$$b = \begin{cases} 1, & \|V_b\| > g \times \sqrt{w^2 + h^2} \\ 0, & otherwise \end{cases} \tag{6.9}$$

where $\| \bullet \|$ is the magnitude of V_b, g is the constant that we found experimentally, and w and h are the width and height of the block.

6.2.1.1.2 Clustering Block Based on Motion Block Attribute of Consecutive Frames

To detect the text with different sizes, we get b in different block sizes. We set the block sizes as 64, 32, 16, and 8. For an 8×8 block, we get four attributes, b_k^1, b_k^2, b_k^3, and b_k^4, which correspond to the block sizes in 64, 32, 16, and 8. For an 8×8 block of the ith frame, we define the motion block feature $M_k^i = b_k^1, b_k^2, b_k^3, b_k^4$, where k indicates the k-th motion block and i is the frame index number. The motion block attribute of consecutive frames mb^k is obtained by the motion block feature M_k^i as $mb^k = \{M_k^i | i = n_b, \dots, n_e\}$, where b_b and n_e are the first and last frame numbers of consecutive frames, respectively.

We clustered the motion blocks into three classes, which refer to three different types of motion perception field. Figure 6.2 illustrates the MPF retrieved by the clustering motion block based on mb^k. We use "+" to indicate the motion

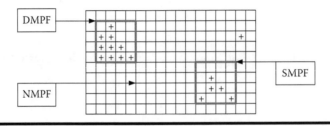

Figure 6.2　Motion perception field illustration.

the novel key frame extraction method using the panoramic frame described in Chapter 6, as well as a representative feature extraction method that we propose. These methods can ensure that we obtain enough representative and robust features for scene recognition. Finally, a model is designed to recognize the scene category of each video clip.

7.2 Related Work

A scene is defined as a site where the image or video is taken. Some examples of scenes include the office and bedroom. So far, there are many related scene recognition methods, and some of them perform well, especially for images. For example, a probabilistic neural network (PNN) was used for indoor versus outdoor scene classification [71]. Zhou et al. [226] presented a novel Gaussianized vector to represent the scene images for unsupervised recognition. Greene and Oliva [68] incorporated the idea of using global frequency with local spatial constraints to recognize scenes.

However, scenes in the same category are presented in various forms of appearances; thus, many researchers have thought that local features are more suitable for constructing the scene model. For example, Liu and Shah [120] utilized a maximization of mutual information (MMI) co-clustering approach to discover clusters of semantic concepts for scene modeling. Li et al. [57] proposed a hierarchical Bayesian approach to classify 13 image scenes. Moreover, Lazebnik et al. [102] argued that methods based on a basic bag-of-features representation do scene recognition better. Thus, they remained sympathetic to the goal of developing robust and geometrically invariant structural object representations and proposed spatial pyramid matching for recognizing natural scene categories. In addition, Wu and Rehg [190] introduced a new visual descriptor, named census transform histogram (CENTRIST), to recognizing both topological places and scene categories. Xiao et al. [192] proposed the extensive Scene understanding (SUN) database that contains 899 categories and 130,519 images for scene recognition.

Nevertheless, the aforementioned methods mainly handle the recognition for specially collected images and cannot adequately deal with the same problems in movies. Recently, there have been several works on movie scene recognition, such as [84] and [158]. Huang et al. [84] developed the scene categorization scheme using a hidden Markov model (HMM)-based classifier. However, these methods concentrated on simple videos only, that is, basketball, football, and commercial videos, and cannot be applied to movies.

Movie scene recognition is more challenging since the physical location appearance depends on various camera viewpoints, partial occlusion, lighting changes, and so on. Schaffalitzky and Zisserman [164] described progress in matching shots of the same three-dimensional location in a film. Both their local invariant descriptor extraction and shot matching process are very time-consuming. Meanwhile, an improved unsupervised classification method was proposed by Héritier et al. [81,154] to extract and link place features and cluster recurrent physical locations (key places) within a movie. Their work focused on near-duplicate detection, which is composed of footage or images of the same object or same background but taken at different times or different places. Bosch et al. [24] also presented a pLSA-based scene classification method mainly for images, but they tested key frames from a movie. In their movies, there are only a few images that could be accurately classified. Thus, it is still a difficult task to directly use an image scene recognition method to do movie scene recognition.

Ni et al. [89] presented an efficient algorithm for recognizing the location of a camera or robot in the learned environment using only the images it captures. However, the location or scene in movies is not as specific and stable. In fact, the works most related to our approach are the studies of [50] and [129]. Engels et al. [50] proposed the automatic annotation scheme for unique locations from videos. Although it was satisfactory and accurately annotated video locations with location words, it was based on the hypothesis that transcripts are available. Marszalek et al. [129] proposed a joint framework for action and scene recognition and also demonstrated the enhanced recognition for both of them in natural video. However, their main purpose was to do action recognition, and their scene recognition results were not satisfactory.

In order to deal with the various appearances in movie scene recognition, we should extract the more robust and representative feature patches and more efficient recognition model, as well as use domain knowledge. Thus, in this chapter, we use the panoramic frame as the key frame and choose representative feature patches (RFPs) to represent each video clip. Meanwhile, local patches dropped in human regions are excluded because human regions are always too noisy for movie scenes. In addition, we consider reoccurring movie scenes that refer to the same place as optimal candidates for enhanced recognition.

7.3 Overview

As shown in Figure 7.1, the proposed method consists of five stages: (1) video segmentation, (2) key frame extraction, (3) representative local feature extraction, (4) latent Dirichlet analysis (LDA)-based classifier construction, and (5) video scene correlation-based enhancement. At first, the video is segmented into shots and scenes; this is introduced in Section 7.4. Then, the panoramic frame is obtained as the key frame for each shot, and representative feature patches are extracted from the panoramic frames belonging to the same video scene, detailed in Section 7.5. After that, in Section 7.6, the Bayesian classifier with an LDA model is trained to recognize the scene category for each video scene, and enhanced scene recognition processing is implemented with near-duplicated video scene detection in Section 7.7. The final processes are described in Section 7.8.

7.4 Video Segmentation

In this section, video segmentation, including shot boundary detection and scene detection, is adopted to segment each movie into clips. There are two different meanings for the word *scene*: one is used in scene recognition, and the other is the description that refers to a group of shots in video composition. We maintain the meaning for *scene* in scene recognition and use video scene (VSC) to represent the video composition definition for a notational distinction. At first, the accelerating shot boundary detection method depicted in Chapter 3 is used to segment a movie into shots efficiently. After that, the proposed multimodality movie scene detection method, using kernel canonical correlation analysis-based feature fusion, is adopted to get video scenes, as shown in Chapter 5.

7.5 Representative Feature Patch Extraction

A VSC usually refers to a group of shots taken in the same physical location; thus a single frame does not obtain sufficient information. Therefore, we get several key frames in a VSC and extract the representative features in those key frames. Concretely, while the frames in a shot are captured by one-time camera motion, we use the panoramic frame obtained by video (frame to frame) registration (depicted in Chapter 4) as the key frame. Then, in order to get more representative features, RFPs are extracted from the key frame.

Although the panoramic frame contains more comprehensive contents of a VSC, different panoramic frames may refer to different view direction appearances. Therefore, this section explains how to extract more representative feature patches from these panoramic frames for final scene recognition.

In film shooting (we mainly discuss scene recognition in movies and teleplays), as the camera moves in various ways, the captured frame also changes. Thus, the global features are unstable, especially when the camera zooms in or out, sweeps, and tilts that result in the scene content change in each frame, and consequently, is not so comprehensive for scene description. Meanwhile, the local feature is more robust and suitable for this scenario, because local features have proved to be more reasonable for semantic analysis and also more robust for sundry variations and occlusions [135]. The combined analysis of all these local features has a more powerful representative ability for a scene. Recently, among the existing local features, scale-invariant feature transform (SIFT) features have proved to be invariant to image scale and rotation and robust to changes in illumination, noise, and minor changes in viewpoint. Therefore, we adopted the classic SIFT feature point extraction method [124] to do local feature extraction and got a 128-dimension feature patch, as depicted in Chapter 2.

Since there are many local patches in each key frame, a scene always has several key frames. Key frames in the same scene describe the scene from different viewpoints, but also have cross-redundancy. Thus, we cluster the reserved patches in one VSC using the *K*-means algorithm and assign the label of the centroid to each cluster. These labeled patches are named representative feature patches.

So far, a movie scene *s* is supposed to be composed of a shot set, that is, $s = \{t_1, , t_n\}$. Meanwhile, a panoramic-based key frame p_i is obtained using the method depicted in Chapter 4 for each shot t_i. Specifically, the process for representative feature patch extraction is depicted as follows:

1. For the panoramic key frame set $P = \{p_1, p_2, , p_n\}$, we extract all the SIFT features or points in each key frame, and fea_i refers to the SIFT points extracted in p_i.

2. Generally, shots are classified into full shot, midshot, and close shot, where full shot has the most abundant scene information, but close shot contains only human or object faces. Thus, in practice, before panoramic synthesis, the human detection operation [25] is applied to the middle frame of each shot at first, and we exclude shots with a large portion (more than half of the frame) of the human region in middle frames directly, which corresponds to the close-up shots and medium close-up shots. Actually, the proposed approach cannot handle the movie or video whose actors are not people, but we can adopt other object detection operations rather than human detection.

3. In addition, human regions always shelter the background; thus, the human regions should be considered as the noise or obstacle for movie scene recognition. As shown in Figure 7.2, for the remaining shots, we extract the panoramic frames as key frames, locate the human regions, and then mask them out. Feature patches within the mask are then filtered out, and we reserve feature patches from the remainder, namely, fea_i in key frame p_i.

Figure 7.2　Human detection in key frames.

4. For all feature point sets $\{fea_1, fea_2, , fea_n\}$, the K-means algorithm is used to get K clusters. Actually, we set K as 0.8 times the average number of feature points in all sets.
5. We assign the label of the centroid to each cluster, and these labeled patches are named representative features patches (RFPs).

7.6 Scene Classification Using Latent Dirichlet Analysis

After we extract the RFPs, we discuss how to use these RFPs to do scene classification and recognition. By accounting for and analyzing most of the movies, without loss generally, we choose the most frequent five scene classes—street, in car, office, restaurant, and bedroom—as our destination scenes. The image scene recognition framework in [57] is adopted for scene recognition for single VSCs. After that, considering that there are so many repeatedly and alternatively appearing VSCs, we also explore the correlations among VSCs and use those correlations to enhance or revise the LDA-based classification result in a single VSC to get the final scene class label for a VSC.

Generally, traditional scene classification methods combine the low-level features of color, texture, and shape with supervised learning algorithms, namely "object first and then scene." Recently, in order to overcome the semantic gap between low-level visual features and high-level semantics and reduce the work of manual labeling, researchers have paid more attention to the intermediate features. The key issue for these methods is how to define, extract, and describe the intermediate semantic features. To this end, we adapt the problems of recent work on image analysis by Blei et al. [21], which was designed to represent and learn document models.

Actually, latent Dirichlet analysis is a three-level Bayesian probability model, including word, theme, and document. The structure of document to theme obeys the Dirichlet distribution, while the structure of theme to word obeys multinomial distribution. LDA is an unsupervised learning model that can be used to recognize the themes in a large-scale document set or corpus. In fact, with the ideas of LDA, in our approach, RFPs are first clustered into different intermediate themes, and then into movie scene classes or categories.

The graphical illustration of LDA is shown in Figure 7.3, where the arrow refers to the dependence relationship. Before discussing modeling, we depict the main variables of the LDA model for movie scene recognition:

1. Patch: A patch x is the basic unit of a VSC, namely, a RFP, where RFP is a patch membership from a dictionary of code words indexed by $\{1, \dots, T\}$. The tth element can be represented by a T-vector x such that $x^t = 1$ and $x^s = 0$ for $s \neq t$. Corresponding to the document, theme, and word in [21],

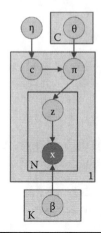

Figure 7.3 LDA theme model.

in this chapter, each VSC corresponds to a document, while a RFP is a word in our model.

2. VSC: A VSC x is a set of N patches without regard to the order, namely, $x = (x_1, x_2, , x_N)$, where x_n is the nth RFPs.

3. Theme: For example, the theme of a camera may be a movie or technology. The theme variable is always the hidden variable; that is, we only know the number of theme variables, but we do not know what the specific theme is.

4. Category or class: A category or class D is a collection of I VSCs denoted by $D = \{x_1, x_2, , x_I\}$.

Now, we describe the LDA model construction and classification with a Bayesian decision.

A VSC s can be generated formally from the model as follows:

1. For each VSC, choose a category label $c \sim p(c|\eta)$, where $c \in \{1, \dots, C\}$, C is the total number of scene categories (five in our approach), and η is a C-dimension vector of a multinomial distribution.

2. Given a particular VSC in category c, we need to draw the parameter that determines the distribution of the intermediate themes, namely, choose $\pi \sim p(\pi|c, \theta)$, where θ is a $C \times K$ matrix and θ_c is a K-dimension Dirichlet parameter conditioned on the category c. K is the total number of themes.

3. For each N patch x_n in the VSC,
   ■ Choose a theme $\ddagger_n$, $\ddagger_n \sim Mult(\pi)$. $\ddagger_n$ is a K-dimension unit vector, where $\ddagger_n^k = 1$ indicates that the kth theme is selected.
   ■ When the theme is defined, choose a patch $x_n \sim p(x_n|\ddagger_n)$, β, where β is a $K \times T$ matrix. K is again the number of themes, and T is the total

number of code words in the codebook. Meanwhile, it is easy to know that $\beta_{kt} = p(x_n^t = 1|{\ddagger}_n^k = 1)$.

Given the parameters θ, η, β, we can get

$$(\mathbf{x}, \mathbf{z}, \pi, \mathbf{c}|\theta, \eta, \beta) = \mathbf{p}(\mathbf{c}|\eta)\mathbf{p}(\pi|\mathbf{c}, \theta) \cdot \prod_{n=1}^{N} \mathbf{p}({\ddagger}_n|\pi)\mathbf{p}(\mathbf{x}_n|{\ddagger}_n, \beta) \qquad (7.1)$$

$$p(\pi|\mathbf{c}, \theta) = \prod_{j=1}^{c} \mathbf{Dir}(\pi|\theta_j)^{\delta(\mathbf{c},\mathbf{j})} \qquad (7.2)$$

$$p({\ddagger}_n|\pi) = \mathbf{Mult}({\ddagger}_n|\pi) \qquad (7.3)$$

$$p(x_n|{\ddagger}_n, \beta) = \prod_{k=1}^{K} \mathbf{p}(\mathbf{x}_n|\beta_\mathbf{k})^{\delta\left({\ddagger}_n^k, 1\right)} \qquad (7.4)$$

Deducing with Laplace estimation, parameter estimation, and Markov chain Monte Carlo (MCMC), the parameter of the model is obtained, and details for that can be seen in [21] and [57].

With the theory from [57], given an unknown VSC s_i, it is represented by a set of code words. We empirically find that the quality of topic distributions is relatively stable if the number of topics is within a reasonable range, and we choose $k = 35$ topics for the construction of LDA. Finally, we have the decision probability of s_i classified to movie scene category c.

$$p(c|s_i, \theta, \beta\eta) \propto \mathbf{p}(\mathbf{s_i}|\mathbf{c}, \theta, \beta)\mathbf{p}(\mathbf{c}|\eta) \propto \mathbf{p}(\mathbf{s_i}|\mathbf{c}, \theta, \beta)$$

where θ, β, and η are parameters learned from the training set and c is the category index. Then, we use a decision probability $p(c|s_i)$ to classify VSC s_i into category c. After that, each VSC is classified into a movie scene category with the largest decision probability.

7.7 Enhanced Recognition Based on VSC Correlation

Actually, there are many VSCs referring to the same physical location in a movie. But the features extracted in different VSCs with different appearances may present various recognition abilities in the recognition model. However, the similarity correlations between these VSCs are easy to obtain, and they are reliable context information for more accurate recognition. Therefore, we use the near-duplicate VSC identification to get the VSC correlations. Given two VSCs s_x and s_y, they are represented with two RFP sets R_x and R_y, respectively. For each RFP in R_x, a series of comparisons is involved to search the nearest neighbor in R_y, and it is computationally expensive. However, the LIP-IS algorithm [221] can be used for faster RFP nearest-neighbor searching. The similarity definition Sim for two RFPs $r_x = \{r(x, 1), r(x, 2), , r(x, 36)\}$ and $r_y = \{r(y, 1), r(y, 2), , r(y, 36)\}$ is defined as

$$Sim(r_x, r_y) = \sum_{i=1}^{3} 6Col(r_{x,i}, r_{y,i}) \tag{7.5}$$

where $0 < Col(r_{x,i}, r_{y,i}) \leq 1$ is the collision function [221].

$$C(r_{x,i}, r_{y,i}) = \begin{cases} 1, & |\mathcal{H}(r_{y,i}) - \mathcal{H}(r_{x,i})| \leq 1 \\ 0, & other \end{cases} \tag{7.6}$$

$\mathcal{H}(r_{x,i})$ is the index of $r_{x,i}$,

$$\mathcal{H}(r_{x,i}) = \left\lfloor \frac{r_{x,i} + 1}{\Delta} \right\rfloor \tag{7.7}$$

Since $-1 \leq r_{x,i} \leq 1$, according to the definition of LIP-IS, it needs to be mapped into eight intervals, and the interval number is used as the index. Therefore, $\Delta = 0.25$.

The matching score of s_x and s_y is summarized in Algorithm 7.1.

Algorithm 7.1 VSC Matching with LIP-IS Filtering Mechanism

Input: VSC s_x and s_y.
Output: Matching score of s_x and s_y.

1: Extract the RFP sets $\mathcal{R}_x$ and $\mathcal{R}_y$ in s_x and s_y, respectively.
2: Hash all RFPs in $\mathcal{R}_x$ to LIP-IS.
3: Initialize $m = 0$.
4: **for** each RFP r_y in $\mathcal{R}_y$ do **do**
5:     Retrieve the nearest neighbor RFP r_x in $\mathcal{R}_x$ to r_y.
6:     If the similarity of r_x and r_y is bigger than T_s, $m + +$.
7: **end for**
8: Given the number of RFPs in $\mathcal{R}_y$ as n, calculate $r = m/n$.
9: Return r as the similarity or matching score of s_x and s_y.

With Algorithm 7.1, we can get the matching score (also named correlation score) $Sc(x, y)$ of s_x and s_y. If $Sc(x, y)$ is bigger than the threshold T (we take $T = 0.68$), they are named the near duplicates of each other. Supposing the near duplicates s_i and s_t are with decision probabilities of $p(c|s_i)$ and $p(c|s_t)$, respectively, and their correlation score is $Sc(i, t)$, we update the decision probability of $p(c|s_i)$ as follows:

$$p'(c|s_i) = \frac{1}{N} \sum_{t=1}^{N} \left(\frac{1}{1 + Sc(i, t)} p(c|s_i) + \frac{Sc(i, t)}{1 + Sc(i, t)} p(c|s_t) \right) \tag{7.8}$$

where N is the number of the near-duplicate VSCs for s_i and $p'(c|s_i)$ is the new decision probability for scene s_i referring to the category c.

According to Equation 7.8, we can see that if the matching score $Sc(i, t)$ is small, it means that the similarity measure is not reliable enough, and the near-duplicate VSC cannot provide credible context information. Otherwise, the two VSCs could be seen as the same location, and their final decision probabilities should be consistent with a weighted-average calculation.

7.8 Experimental Results

In order to evaluate the validity of the proposed movie scene recognition approach, we test the approach on both the public dataset and the dataset collected by us. We adopt the five frequently used movie scenes to do movie scene recognition: street, office, restaurant, bedroom, and in car. Meanwhile, Marszalek et al. [129] have provided a dataset HOLLYWOOD2 (MD) with 12 classes of human actions and 10 classes of scenes (EXT-House, EXT-Road, INT-Bedroom, INT-Car, INT-Hotel, INT-Kitchen, INT-LivingRoom, INT-Office, INT-Restaurant, INT-Shop) with 570 training samples and 582 testing samples, using the alignment of scripts and videos. In order to do reasonable comparison, we map the above 10 scene classes into our previously defined 5 classes to form the public dataset. Especially, EXT-Road will be mapped into street and in car, respectively. Actually, some of the clips in HOLLYWOOD2 are not complete video scenes, and the resolution and brightness are not too good. Thus, we also collect a more suitable dataset (OM Dataset [OD]) manually: street (45 clips), office (58 clips), restaurant (46 clips), bedroom (51 clips), and in car (61 clips). Video clips in OD are VSCs with obvious scene appearances.

7.8.1 Comparison of Different Methods

Since there are not many movie scene recognition methods, we consider the typical video scene recognition method proposed by Marszalek et al. [129], as well as the well-performed image scene recognition method named CENTRIST [190] ($CENTRIST_M$). Meanwhile, we performed these two methods as well as the proposed approach on both the OM dataset and subset of the MD dataset (selecting 10 clips in each category). Since the CENTRIST method is an image-based scene recognition method, in order to be used in video clips in both datasets, we directly chose the middle frame of each shot as the input for this method, and the scene category of the frame with the highest classification score was used as the output. Table 7.1 shows the comparison results.

Table 7.1 Comparison of Proposed Approach with *CENTRIST*$_M$ [190] and *MARCIN*$_M$ [129]

	OUR_M		$MARCIN_M$		$CENTRIST_M$	
	MD	OD	MD	OD	MD	OD
Street	0.81	0.84	0.52	0.55	0.62	0.66
Office	0.74	0.80	0.62	0.63	0.60	0.61
Restaurant	0.61	0.67	0.33	0.40	–	–
Bedroom	0.72	0.71	0.51	0.51	0.69	0.70
In car	0.64	0.64	0.66	0.67	–	–

7.8.2 Performance Evaluation on Using Panoramic Frames

Static scene images are always landscape images and contain more comprehensive features of the whole scene, but the focus or attention in movie scenes is mostly the moving objects. Thus, there are only a few features about scenes in some shots, and even worse, the background referring to the scene is blurred as a special cinematography. However, using panoramic frame-based key frames, we can collect more comprehensive scene features in consequent frames taking into account that redundancy is also abundance. Finally, as shown in Figure 7.4, we get a more efficient

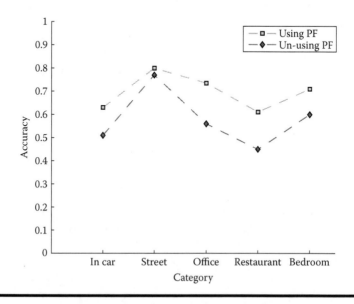

Figure 7.4 Comparisons using or not using panoramic frames.

Figure 7.5 **Confusion matrix of our movie scene dataset.**

recognition performance that archives an improvement of 9% accuracy on average in the five scene categories.

The confusion matrix in all categories is shown in Figure 7.5. In fact, the most confusing pairs of the five categories are office/restaurant, office/bedroom, and restaurant/bedroom, especially for normal key frame extraction. That is because in partial scenes, these pairs of categories share very similar scenes or backgrounds. However, from Figure 7.5, we can see that our approach has satisfactory results in these pairs because different categories present distinguishing appearances in the panoramic frames.

7.8.3 Comparison of Different Experimental Conditions

In order to assess the effectiveness of the feature extraction on nonhuman regions as well as the VSC correlation-based enhanced recognition, we perform several experiments in different experimental conditions. We evaluate the performances of the RFP extraction, removing the human regions, and VSC correlation-based enhanced recognition. The human regions always take a large portion in most frames, but they are noise for scene recognition, which mainly refers to background information. Thus, the human region removing process is introduced to exclude local features in human regions, and the improved recognition accuracy is shown in Figure 7.6.

In Figure 7.7, we compare the performance of the direct SIFT feature extraction–based scene recognition (DSFSR) method, which is the method used in [129], with those of our RFP extraction and human region removal (RFEHRSR) and RFEHRSR combined with the VSC–enhanced recognition method (VERFHRSR). We find

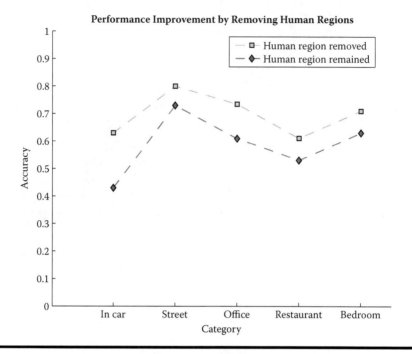

Figure 7.6 Performance improvement with human region removal.

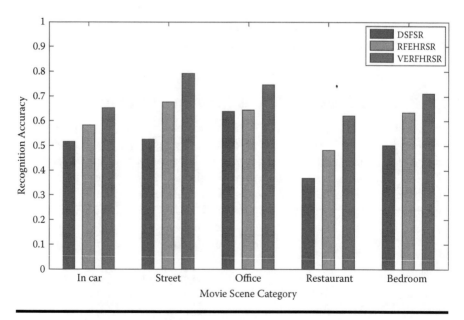

Figure 7.7 Recognition performances in different conditions.

that RFP and human region removal enhance the recognition accuracy. Because we reduce most of the redundant features that are identified as noise by RFPs and remove the human regions, the pure location regions are more efficiently used for training and learning.

Meanwhile, the recognition of outdoor street movie scenes reached a high accuracy of 80%, while the recognition results of three indoor movie scenes were not as good as those of the street scenes. The in-car movie scene seems to be discriminated easily, but the recognition result was not very good; the reason may be that the human region occupies a very large portion in in-car movie scenes, and the number of features used for training and recognition is very small. We conclude that without removing the human regions, the performance decreased obviously, especially in the inside movie scenes of bedroom and office. The VSC correlation enhances the recognition results of individual VSCs. Our approach achieves a satisfactory performance on the five movie scene categories, as shown in Figure 7.7.

7.9 Summary

In this chapter, we studied how to effectively recognize scenes in a movie. More specifically, the movie was efficiently segmented into clips. Then, by stitching the panoramic frame as key frames, we extracted the representative local features in the key frames and further removed the noise of human regions. In addition, during the process of extracting all the local patches in each VSC, the RFPs were chosen to represent the VSC. After that, a LDA-based movie scene recognition model was built by training the collected VSCs. Finally, when the recognition results for each individual VSC were ready, the correlations of VSCs were taken into account for enhanced recognition. Although the recognition results were not dramatically improved, it was a meaningful idea to collectively use both key frame information and related VSC information for video content analysis.

Chapter 8

"Who" Entity: Character Identification

Generally, characters and their actions are the most crucial clues in movies and teleplays. Therefore, face detection, character identification, human tracking, and so on, become very important in content-based movie and teleplay analysis and retrieval. Meanwhile, the characters are often the most important contents to be indexed in a feature-length movie; thus, character identification becomes a critical step in film semantic analysis, and it is also reasonable to use character names as cataloguing items. In this chapter, we describe how to recognize characters in movies and teleplays, and then use character names as cataloguing items for intelligent catalogues.

8.1 Preface

Movies and teleplays are the most popular consuming videos in our daily lives. In addition, with the rapid development of the Internet, television videos not only exist in the traditional broadcast television system, but also appear on the Internet. In recent years, the number of video resources has grown very rapidly, but tackling this large number of video resources is still a hard problem for video resource managers and consumers (the audience), who have to figure out how to effectively manage and properly provide content of user interest, as well as retrieve resources. In addition, these videos are always expanding the story surrounding their characters and their events. Therefore, if we are able to detect and identify the location or fragment where the character appears, and the names of those characters are labeled to these fragments, then they become available for us to do effective and intelligent movie and teleplay management and retrieval. For example, with the recognized characters, the labeled name can be used for actor-based video cataloguing. That is, the video clips corresponding to the same actor can be catalogued to the same directory. Similarly, through video clips with the labeled actors' names, users can search for videos or clips with actors or characters of interest according to their needs, especially for on-demand movies and teleplays.

Generally speaking, character retrieval or actor labeling methods should be archived automatically or semiautomatically with various modality information (such as images, videos, audio, and text). This should be implemented as far as possible with an automated mode rather than manual intervention. In fact, character recognition or name labeling is realized with face detection and recognition. Therefore, the efficiency and accuracy of the face acquisition mode and face recognition method are crucial criteria for character retrieval and actor labeling.

On the one hand, a video is composed of a large number of frames (images), so mature face detection and recognition technologies based on images, can be naturally adopted and extended to video character detection and identification. However, in the videos, environmental factors (e.g., light and weather) and video quality (picture clarity and resolution) changes directly affect the face detection and recognition performance. In addition, the same person will also face different angles and positions, which causes difficulty in identifying and labeling characters in videos.

On the other hand, unlike traditional surveillance videos, one of the greatest features of movies and teleplays is that the camera is not fixed; it moves corresponding to the changes of character or scene content. The movement of both camera and characters action makes the faces in videos more complex and difficult to detect. Consequently, this results in more challenges and difficulties in character identification and actor labeling in videos.

In this chapter, we propose a novel semisupervised learning strategy to address the problem of character identification, especially celebrity identification. The video context information is explored to facilitate the learning process based on the assumption that faces in the same video track share the same identity. Once a frame within a track is confidently recognized, the label can be propagated through the whole track, referred to as the *confident track*. More specifically, given a few static images and vast face videos, an initial weak classifier is trained and gradually evolves by iteratively promoting the confident tracks into the labeled set. The iterative selection process enriches the diversity of the labeled set such that the performance of the classifier is gradually improved. This learning theme may suffer from semantic drifting caused by errors in selecting the confident tracks. To address this issue, we propose treating the selected frames as *related samples*—an intermediate state between labeled and unlabeled, instead of labeled as in the traditional approach. To evaluate the performance, we construct a new dataset that includes 3000 static images and 2700 face tracks of 30 celebrities. Comprehensive evaluations on this dataset and a public video dataset indicate significant improvement of our approach over established baseline methods.

8.2 Introduction

Recently, automatic character identification [10, 54, 163], which detects character faces in photos or movies and associates them with corresponding names, has attracted a lot of attention in the computer vision field. Among the many applications of character identification, celebrity-related tasks draw the most attention due to the common interest of people in celebrities. Furthermore, celebrity identification has been considered a crucial step for image and video semantic analysis [12, 19, 163], with growing research enthusiasm in multimedia technologies.

To this end, researchers have proposed many methods for celebrity identification [17, 217, 220]. Nevertheless, as mentioned in [10], the problem still remains tremendously challenging due to: (1) a lack of precisely labeled training data; (2) significant visual variations in terms of human poses, light, facial expressions, and so on; and (3) low resolution, occlusion, nonrigid deformation, large motion blur, and complex backgrounds in realistic photographic conditions.

An intuitive way to deal with these challenges is to collect a large-scale face database with sufficient data diversity and a reliable ground-truth label. However, the enormous amount of manual work required in data labeling hinders constructing such a dataset. On the other hand, the rapid development of the Internet provides easy access to a large collection of unlabeled face data. Commercial search engines, such as Google, can return a large pool of images corresponding to a certain celebrity within several milliseconds. Large video sharing platforms such as YouTube receive around 100 h of videos uploaded every minute. The massive data available online and the easy accessibility have motivated researchers to investigate how to improve the performance of traditional learning-based multimedia analysis methods utilizing such a large unlabeled dataset. As a result, semisupervised learning [15, 225, 228] has drawn plenty of research interest during the past few decades.

In this work, we propose a novel way to utilize video context to boost recognition accuracy for celebrity identification with limited labeled training images. Compared with face images returned by search engines, faces in videos are captured in an unconstrained way and present more variations in pose, illumination, and so forth. Moreover, although noisy, videos are usually accompanied by reliable context information that can be used for de-noising. In this chapter, we extract face tracks from downloaded videos and build the celebrity identification framework with a simple but effective assumption: faces from the same face track belong to the same celebrity.

More specifically, our system learns a weak classifier from a few labeled static images. The learned classifier is then used to predict the labels and confidence scores of all the frames within each video track. The frames are ranked with regard to the confidence scores, and the track possessing the frame with the highest confidence score is chosen as the confident track. The video constraint enables the propagation of predication labels across the frames of the confident track, which is then promoted into the *related set*, as illustrated in Figure 8.1. The update of the classifier is realized under the supervision of related samples in the related set. This select update process

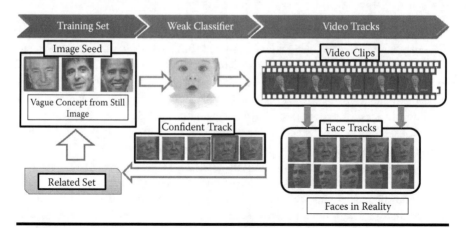

Figure 8.1 Illustration of the proposed adaptive learning framework. The initial classifier is trained on a small set of static images (image seeds) and then used to label the frames within each video track. If a certain frame is assigned with a confident label, all the frames within the same track are promoted into the related set and utilized to update the classifier in the next iteration such that the classifier gradually evolves.

is iterated multiple times such that the classifier evolves with improved discriminative capacity gradually. The proposed learning theme has a certain analogy to some recent biological studies of the cognitive process of human brains. According to adaptive resonance theory (ART) [69], the human brain forms the resonant state depicting the links between visual inputs and semantics in the initial learning stage and searches for good enough matches to enhance the understanding of objects or people gradually in an adaptive learning manner.

The proposed method shares a certain similarity to self-training [131, 203] since both adopt a mechanism of iteratively selecting samples from the unlabeled set to improve the performance. The difference lies in that our approach introduces the video context constraint into the selection process, such that the positive samples that cannot be recognized confidently may still be promoted. Self-training often suffers from the well-known semantic drifting [168]. It occurs when the size of the labeled set is too small to constrain the learning process. More specifically, the errors in selecting the best samples may accumulate, and consequently, newly added examples tend to stray away from the original concept. Existing solutions to semantic drifting mainly focus on improving the accuracy in the selection process, among which co-training and active learning are two major research directions. This chapter, on the contrary, explores from a different perspective. Instead of struggling to select the correct samples, we aim to design a classifier robust to the selection errors by treating selected samples as related rather than labeled. More specifically, we decrease the influence of the selected samples, or related samples, as termed in this work, to guarantee that their influence is weaker than that of the labeled samples.

Furthermore, the influence of a specific related sample is reweighted based on the corresponding confidence score, so that discriminative samples are emphasized, while noisy, and nondiscriminative samples are suppressed at the same time.

8.3 Related Work

Celebrity identification is a specific application of face recognition. Previous works on celebrity identification can be generally categorized into two groups: (1) face recognition considering correspondence between face and text information and (2) face recognition utilizing a large manually labeled image or video training set.

In the first group, the textual information is used to provide an extra constraint in the learning process. An early work of Satoh et al. introduced a system to associate names located in the sound track with faces. Berg et al. [17] built up a large dataset by crawling news images and corresponding captions from Yahoo! News. Everingham et al. [54] explored textual information in scripts and subtitles and matched it with faces detected in TV episodes. However, the main disadvantage in the studies of the first group is the heavy dependence on associated textual information. In most cases, the assumption that textual information is available does not always hold, and errors may occur in the given text description.

The other group aims to learn a discriminative model based on a manually labeled dataset. For example, Tapaswi et al. [178] presented a probabilistic method for identifying characters in teleplays or movies, and the face and speaker models were trained on several episodes with manual labeling. In the work of Liu and Wang [123], a multicue approach combining facial features and speaker voice models was proposed for major cast detection. However, the performance of supervised learning methods mentioned above was usually constrained by the insufficiency of labeled training samples. Much research interest has been drawn from scenarios where only a limited number of labeled training samples were available, which is much more common in reality.

To overcome this scarcity in the training set, semisupervised learning (SSL)–based methods [15,225,228] are proposed in many studies based on the assumption that unlabeled data contain the information of underlying distribution and thus can facilitate the learning process. The explosive development of video sharing websites, such as YouTube, provides easy access to such a large unconstrained and unlabeled training set. Plentiful studies have been conducted using video data in multiple active fields of computer vision, including object detection [149,202], object classification [200], person identification [14], action recognition [31], and attribute learning [38].

Among various SSL methods, one of the classic ones is the bootstrapping-based method, also known as self-training. For instance, Cherniavsky et al. [36] trained a classifier on a set of static images and used it to recognize attributes in videos. Chen and Grauman [31] addressed the action recognition task by learning generic body motion from unconstrained videos. In their example-based strategy, the most confident pose is located in a nearest-neighbor manner and then added into the training set. Kuettel et al. [98] proposed a segmentation framework on the ImageNet dataset by recursively exploiting images segmented so far to guide the segmentation of new images in a bootstrapping manner. Choi et al. [38] proposed expanding the visual coverage of training sets by learning from confident attributes of unlabeled samples. It was also claimed that even though some attributes were selected from other categories, they could lead to improvement in category recognition accuracy.

A typical issue in the self-training methods is caused by the error in labeling confident samples in each iteration—early errors will accumulate by including more and more false positive samples, causing semantic drifting, as mentioned in [168]. Most researchers solve this problem by trying to increase the labeling accuracy in selection. Conventional approaches include co-training [55] and active learning [22]. Active learning iteratively queries the supervision of the users on the least certain samples. Li and Guo [108] proposed an adaptive active learning method by introducing a combined uncertainty measurement. They selected the most uncertain samples to query users' supervision. These selected samples are added into the training set and used to retrain the classifier. Co-training or multiview learning, on the other hand, learns a classifier on several independent feature sets or views of data [22] or learns several different classifiers from the same dataset [64]. Saffari et al. [162] proposed a multiclass multiview learning algorithm, which utilized the posterior estimation of one view as a prior for classification in other views. In [137], Minh et al. introduced reproducing Kernel Hilbert Space (RKHS) of vector-valued functions into manifold regularization and multiview learning and achieved state-of-the-art performance.

Incremental learning or online learning [65, 66] also includes a mechanism of iteratively updating the classifier. A common assumption is that the training samples with labels are given in a streaming manner; that is, not all the training samples are presented at the same time. Incremental learning cannot select the confident unlabeled data as in self-training, and its performance is quite sensitive to the label noises. In this work, we focus on learning a robust classifier with noisy selected samples. Thus, incremental learning is out of scope in this work.

We propose an adaptive learning approach for celebrity identification by incorporating the video context information. Moreover, we introduce the concept of related sample to address the problem of semantic drifting. Instead of struggling to prevent the error in labeling unknown samples, we aim to obtain a classifier that is robust to selection errors such that the performance can be improved steadily.

8.4 Overview of Adaptive Learning

Adaptive resonance theory (ART) [69] is a cognitive and neural theory to describe how the brain learns to categorize in an adaptive manner. According to ART, the human brain initializes the resonant states, linking the visual inputs to semantics via "supervised learning," and then tries to find good enough matches for the concepts in everyday life. These matches are then used for updating the resonant states in the learning process.

According to ART, a baby may learn in a two-stage manner: initial learning and adaptive learning.

■ Initial learning. A newborn baby does not have much knowledge, that is, resonant states, of recognizing a certain object or person. Parents, acting as supervisors, show the baby the links between words (labels) and visual information and provide some initial labeled samples.

■ Adaptive learning. The baby observes the world by himself or herself. When a certain status of a person matches with the initial pictures in the brain (good match), the baby connects all the visual information of this person with the existing knowledge to update.

Sharing a similar spirit, our framework includes a two-stage learning mechanism on a training dataset consisting of (1) labeled images for initial learning and (2) unlabeled noisy data from the Internet for adaptive learning. The images are retrieved from Google Image using the name of each celebrity as the query word and then manually labeled. For collecting the noisy data, we download video clips from YouTube with tags relevant to each celebrity. Faces in the static images online are usually taken under similar conditions, for example, similar pose, facial expression, and illumination. However, faces in the videos present more variations and thus provide more diverse training samples for adaptive learning. Note that the collected videos are noisy due to (1) the videos not being relevant to the celebrity and wrongly selected because of tagging errors of the users and (2) videos containing several individuals. Thus, such videos are treated as unlabeled data and fed into the classifier without using the ground-truth identification during training.

In this chapter, we extract multiple face tracks from the collected videos and exploit the video context information within the face tracks. We introduce the video constraint into the adaptive learning process; that is, faces from the same track belong to the same identity. The video constraint has a natural connection with the baby learning process. The visual perception of the baby is continuous, and the baby is able to differentiate the correspondence between the consecutive frames, that is, whether these frames share the same identity. Namely, the baby organizes the visual perceptions in the real world as tracks of consecutive frames that belong to the same identity. Our proposed video constraint possesses a similar spirit.

Before introducing the details of our methods, we define some notation here for formal description. Suppose we are given in total n training samples of N individuals, which include l labeled samples and u unlabeled samples (video tracks), that is, $n = l + u$. We denote the initial labeled image set as $\mathcal{L}_o = \{(x_1, y_1), \dots, (x_l, y_l)\}$, where y_i represents the label for the sample x_i. The unlabeled video set consists of K face tracks $\{T_i \mid i \in \{1, 2, \dots, K\}\}$ with $K \leq u$ and is denoted as $\mathcal{U} = \{(x_{l+1}, \dots, x_{l+u})\}$. Here $\{x_i, i = l + 1, \dots, n\}$ are extracted frames from the face tracks.

The most straightforward way of utilizing the unlabeled samples is to treat the most confident unlabeled track T_i as labeled based on the corresponding confidence score. Here the confidence score can be computed based on the classifier learned from a few labeled samples. These tracks are termed confident tracks, which correspond to the good enough match in the baby learning process [69]. All the frames within are then assigned with the same label as the most confident frame and promoted into the related set, denoted as $\mathcal{L}_r$ (details are in Section 8.5). Afterwards, the classifier is retrained with the current labeled set, the union of the initial labeled set and the discovered related set, $\mathcal{L} = \mathcal{L}_o \cup \mathcal{L}_r$. The updated classifier then predicts the labels of all the remaining frames in the video set. To identify multiple celebrities, the classifier is trained in a one versus all manner. More specifically, we train N binary classifiers, each of which is learned by taking one class of samples as positive and the remaining $N - 1$ classes of samples as negative. The most confident tracks are then selected per class in each iteration. This is to avoid the dominance of a certain class in the track selection and balance the response magnitude of all the classifiers. The confidence score of each frame belonging to class j is computed via a soft-max function $g_j(\cdot)$ on the response of each classifier:

$$g_j(x_i) = \frac{\exp\{f_j(x_i)/\eta\}}{\sum_k \exp\{f_k(x_i)/\eta\}} \tag{8.1}$$

where $f_j(\cdot)$ denotes the binary classifier for the class j and η is a trade-off parameter for approximating the max function. Large η renders almost the same scores for different inputs, while small η enlarges the gaps among the output confidence scores.

We compute the confidence scores of all the frames within each face track. The maximum of these confidence scores within each track is denoted as MaxF, and the minimum is denoted as MinF. Different face tracks are ranked in terms of their MaxF scores, and only the top S tracks are selected as candidates for the following selection. The candidate tracks are then ranked in terms of their MinF scores, and the track with the largest MinF score is selected as the confident track. This selection process is graphically illustrated in Figure 8.2. Using this mechanism, we aim to choose the track in which a certain frame is recognized as the best match, and the rest frames are considered to be good enough matches. For better understanding of this proposed mechanism, consider an extreme case where there are a large number of candidate tracks. For this case, we actually select the most confident tracks by the averaged confidence scores of all the tracks. However, the selection results for this setting are possibly the tracks with minor between-frame variation. This may limit the generalization performance of the learned classifier. On the other hand, if S is too small, for example, $S = 1$, it is quite likely to include false tracks, especially when the initial classifier is trained on a small labeled set. Considering the total number of video tracks (around 2700) in our experiments, we empirically set $S = 5$ throughout the experiments. This small value of S may achieve a good trade-off between the diversity of the chosen tracks and the selection accuracy. The framework of adaptive learning based on such a selection strategy is described in Algorithm 8.1.

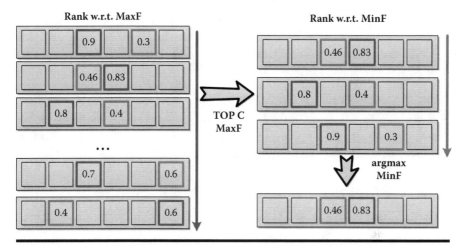

Figure 8.2 Illustration of confident track's selection mechanism. Each large block represents a face track. The small red block refers to the most confident track, and the blue block refers to the least confident track. Their corresponding confidence scores are shown inside. The first selection step (left) is based on MaxF, and the second step (right) is based on MinF.

Algorithm 8.1 Framework of Adaptive Learning

Input:

Initial labeled set $\mathcal{L}_o$, related set $\mathcal{L}_r = \emptyset$, unlabeled set $\mathcal{U}$, number of classes N, maximal iteration number N_{iter}, and S for TOP-S setting.

Output: Final classifier $F = \{f_1 \ldots, f_N\}$.

1: **for** $i = 1 : N_{iter}$ **do**
2:     $\mathcal{L} \leftarrow \mathcal{L}_o \cup \mathcal{L}_r$
3:     Train classifier $F^{(i)} = \{f_1^{(i)}, \ldots, f_N^{(i)}\}$ on $\mathcal{L} \cup \mathcal{U}$.
4:     Compute $g_k(x_j)$, $\forall x_j \in \mathcal{U}$, $k = \{1, \ldots, N\}$.
5:     **for** $k = 1 : N$ **do**
6:        Compute $MaxF$ for each track.
7:        Choose top S tracks as candidates according to $MaxF$.
8:        Select track $\mathcal{T}_p$ with max $MinF$ from S candidates.
9:        Set labels for $x_j \in \mathcal{T}_p$ as k.
10:        $\mathcal{L}_r \leftarrow \mathcal{L}_r \cup \{x_j \in \mathcal{T}_p\}$
11:        $\mathcal{U} \leftarrow \mathcal{U} \setminus \{x_j \in \mathcal{T}_p\}$
12:     **end for**
13: **end for**

In general, adaptive learning is more robust to various changes in terms of pose, facial expression, and so forth. Unlike traditional semisupervised learning, confident samples in adaptive learning obtain much higher influence than the remaining unlabeled samples in the next iteration of training. With the introduced video constraint, the labels are propagated from confident frames to those frames that are difficult to label based on information from the limited initial image seeds. The promoted unconfident frames usually contain faces with more variations than the initial labeled samples. As a result, the classifier is trained with enriched labeled data with high diversity, and thus gains improvement on its generalization performance.

8.5 Adaptive Learning with Related Samples

The aforementioned straightforward adaptive approach simply treats *related samples* exactly the same as labeled samples in $\mathcal{L}_o$. Such an approach only works in the ideal case where no errors occur in selecting the confident tracks. However, selection errors are generally inevitable for the following two reasons: (1) poor discriminative capability of the learned classifier in the initial learning stage where the classifier is trained only with a small number of labeled images and (2) high similarity between different persons in certain frames. The errors in the selection process will cause semantic drifting [168] and degrade the performance of the classifier. To address this problem, we introduce the concept of related samples, which is a comprise between labeled and unlabeled samples. Selected related samples are given higher weights than the remaining unlabeled samples, but lower weights than the initial labeled samples in training the classifier. As a result, the initial accurately labeled data still contribute most to the learning process such that the undesired semantic drifting effect brought about by promoting related samples is alleviated in a controlled manner. In the following subsections, we briefly review the Laplacian support vector machine (LapSVM) for semisupervised learning, and then introduce our proposed related LapSVM, which integrates the adaptive learning and related samples.

8.5.1 Review of LapSVM

We formulate the aforementioned ideas under the generalized manifold learning framework. In particular, we adopt Laplacian SVM, introduced by Belkin et al. [15], as a concrete classifier learning method in this work. In this section, we first give a brief review of LapSVM.

LapSVM is a graph-based semisupervised learning method. A sample affinity graph is denoted as $\mathcal{G} = \{V, E\}$, where V represents the set of nodes (data samples) and E refers to edges whose weights specify pair-wise similarity defined as follows:

$$W_{ij} = \exp(-\|x_i - x_j\|^2 / 2\sigma^2) \tag{8.2}$$

where σ is a parameter controlling the similarity based on sample Euclidean distance and is determined via cross-validation in this work.

In LapSVM [15], classifier f is learned by minimizing the following objective function:

$$J(f) = \sum_{i=1}^{l} \max(1 - y_i f(x_i), 0) + \gamma_A \| f \|_A^2 + \gamma_I \| f \|_I^2 \qquad (8.3)$$

where $\| f \|_A^2$ represents the regularization in the corresponding reproducing kernel Hilbert space (RKHS) to avoid overfitting. $\| f \|_I^2$ embodies the smoothness assumption on the underlying manifold; that is, samples with high similarity have similar classifier responses. Here, we adopt a graph-based manifold regularizer as $\| f \|_I^2 = \sum_{i,j} (f(x_i) - f(x_j))^2 W_{ij}$.

By defining the classifier in the RKHS according to the representer theorem [165], we have the following classifier representation:

$$f(\cdot) = \sum_{i=1}^{l+u} \alpha_i k(x_i, \cdot) \qquad (8.4)$$

where $k(x_i, \cdot)$ is a kernel function in RKHS. In this work, we adopt linear kernel, trading off the performance and computational complexity, that is, $k(x_i, x_j) = x_i x_j$.

By substituting Equation 8.4 into Equation 8.3, the objective function is equivalently rewritten as

$$J(\alpha) = \sum_{i=1}^{l} \max(1 - y_i f(x_i), 0) + \gamma_A \alpha^T K \alpha + \gamma_I \alpha^T K L K \alpha \qquad (8.5)$$

where $\alpha = \{\alpha_1, \alpha_2, \dots, \alpha_n\}^T$ and K is the $n \times n$ gram matrix over labeled and unlabeled sample points. $L = D - W$ is the Laplacian matrix on the adjacency graph $\mathcal{G}$, where D is the diagonal matrix with $d_{ii} = \sum_j w_{ij}$ and W is the weight matrix defined in Equation 8.2.

LapSVM can be directly applied in our adaptive learning framework. However, as pointed out before, the cumulative error in labeling the unlabeled data may cause the problem of semantic drifting. In the following section, we introduce the proposed related LapSVM to solve the problem.

8.5.2 Related LapSVM

Intuitively, to solve the problem of incorrect sample selection, the influence of selected samples should be more significant than that of the remaining unlabeled samples, but not greater than that of the initial original labeled samples. Referring to the LapSVM [15], labeled data are prone to be the support vectors or, in other words, lying on the margin such that $y_i (w^T x_i + b) = 1$, while there is no such constraint on unlabeled data. Selected frames, however, should lie between the decision boundary

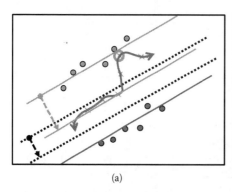
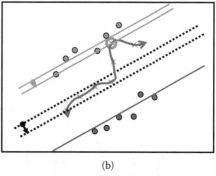

(a)   (b)

Figure 8.3 Illustration of naive adaptive learning and related LapSVM. Blue and red dots represent labeled samples for positive and negative classes, respectively. Yellow stars represent face frames in a face track (gray curve). A certain frame (star in blue circle) is recognized as the most confident sample with a positive predicted label. Block (a) shows the change of margin (blue and red line) and decision boundary (black dashed line), as indicated by the colored arrows, for naive adaptive learning. Block (b) shows the change after including the concept of related sample. For naive adaptive learning, the margin is completely determined by selected samples; that is, the initial labeled images are unable to constrain the learning process. However, for related LapSVM, the influence of related samples does not overtake the original labeled set and the margin is retained as desired.

(uncertain unlabeled data) and the margin (labeled data). By considering the hard constraint in the video, frames from the same track should be put on the same half space with respect to the classifier decision boundary, as shown in Figure 8.3. These selected samples are treated as related samples, lying between the labeled and unlabeled samples.

We propose the related LapSVM to incorporate the concept of related sample into LapSVM. Formally, via introducing a weight ρ for the related samples in deciding the classifier boundary, the objective function of LapSVM in Equation 8.5 is changed into

$$\mathcal{J}(\varepsilon, \alpha) = \sum_{i=1}^{l} \varepsilon_i + \gamma_A \alpha^T K \alpha + \gamma_I \alpha^T K L K \alpha$$

$$\text{s.t. } y_i \left(\sum_{j=1}^{l+u} \alpha_j k(x_j, x_i) + b \right) \geq 1 - \varepsilon_i, \forall x_i \in \mathcal{L}_o$$

$$y_T^i \left(\sum_{j=1}^{l+u} \alpha_j k(x_j, x_i) + b \right) \geq (\rho \cdot C_T^i) - \varepsilon_i, \forall x_i \in \mathcal{L}_r$$

$$\varepsilon_i \geq 0, \forall x_i \in \mathcal{L}, \ 0 \leq \rho \leq 1 \tag{8.6}$$

where ε_i is the slack variable for x_i. The predicted label y_T^i and confidence score $\mathcal{C}_T^i$ for the most confident frame in track T_i are defined as follows:

$$\mathcal{C}_T^i = \max_{x_j \in T_i} g(x_j)$$

$$y_T^i = \mathrm{sgn}\left(f\left(\arg\max_{x_j \in T_i} g(x_j) \right) \right) \tag{8.7}$$

where $g(\cdot)$ is the soft-max function for calculating the confidence score. With Equation 8.7, each face track is tagged with the same label as the most confident sample within.

As shown in Equation 8.6, each related sample $x_i \in \mathcal{L}_r$ is placed on a hyperplane with a distance $\rho \cdot \mathcal{C}_T^i$ to the decision boundary. The farther the hyperplane lies from the decision boundary, the greater the influence the related samples lying on it will have in defining the decision boundary. This is based on the assumption that the track with the sample of a higher confidence score has a higher probability to be the correct track, and thus should have a stronger constraint in the training phase. The constraint in Equation 8.6 guarantees that the influence of a certain related sample is proportional to the corresponding confidence score. Also, a slack variable is imposed for each related sample, similar to the soft-margin concept in traditional SVM. ρ is a parameter in the range [0, 1] to control the upper bound of the margin for related samples. A larger ρ indicates a stronger constraint on related samples. When ρ is set to 0, we only require all the frames within the same track to lie on the same half space of the decision boundary.

Following an optimization method similar to that in [15], the problem in Equation 8.6 can be written in the following Lagrange form:

$$Lg(\alpha, \varepsilon, b, \beta, \lambda) = \sum_{i=1}^{l} \varepsilon_i + \gamma_A \alpha^T K \alpha + \gamma_I \alpha^T K L K \alpha$$

$$- \sum_{\forall i, x_i \in \mathcal{L}_o} \beta_i \left(y_i \left(\sum_{j=1}^{l+u} \alpha_j k(x_j, x_i) + b \right) - 1 + \varepsilon_i \right) - \sum_{i=1}^{l} \lambda_i \varepsilon_i$$

$$- \sum_{\forall i, x_i \in \mathcal{L}_r} \beta_i \left(y_T^i \left(\sum_{j=1}^{l+u} \alpha_j k(x_j, x_i) + b \right) - \rho \cdot \mathcal{C}_T^i + \varepsilon_i \right) \tag{8.8}$$

According to the Karush-Kuhn-Tucker (KKT) conditions, we set the derivatives of L_g in terms of b and ε_i as zeros, which yields

$$\frac{\partial L_g}{\partial b} = 0 \Rightarrow \sum_{i, x_i \in \mathcal{L}_o} \beta_i y_i + \sum_{i, x_i \in \mathcal{L}_r} \beta_i y_T^i = 0,$$

$$\frac{\partial L_g}{\partial \varepsilon_i} = 0 \Rightarrow 1 - \beta_i - \lambda_i = 0 \Rightarrow 0 \leqslant \beta_i \leqslant 1 \tag{8.9}$$

By substituting Equation 8.9 into Equation 8.8 and canceling b, λ, ε, the Lagrangian function becomes

$$L_g(\alpha, \beta) = \gamma_A \alpha^T K \alpha + \gamma_I \alpha^T K L K \alpha$$

$$-\alpha^T K J_L^T Y \beta + \sum_{\forall i, x_i \in \mathcal{L}_o} \beta_i + \sum_{\forall i, x_i \in \mathcal{L}_r} (\rho \cdot C_T^i) \beta_i \qquad (8.10)$$

$$\text{s.t.} \quad 0 \leq \beta_i \leq 1, \forall x_i \in \mathcal{L}_o \cup \mathcal{L}_r$$

Here Y is a diagonal-labeled matrix whose nonzero entries are set as label y_i for samples in $\mathcal{L}_o$ or predicted label y_T^i for samples in $\mathcal{L}_r$; we also define $J_L = [I \quad 0]$, where I is an identity matrix with a size equal to the cardinality of set $||\mathcal{L}||$.

Applying the KKT conditions again, we represent α by β:

$$\frac{\partial L_g}{\partial \alpha} = 0 \rightarrow \alpha = (2\gamma_A I + 2\gamma_I L K)^{-1} J_L^T Y \beta \qquad (8.11)$$

and K is invertible since it is positive semidefinite.

Finally, the corresponding dual form of Equation 8.6 can be rewritten as follows:

$$\max_{\beta} \quad \sum_{\forall i, x_i \in \mathcal{L}_o} \beta_i + \sum_{\forall i, x_i \in \mathcal{L}_r} (\rho \cdot C_T^i) \beta_i - \frac{1}{2}\beta^T Q \beta$$

$$\text{s.t.} \quad \sum_{\forall i, x_i \in \mathcal{L}_o} \beta_i y_i + \sum_{\forall i, x_i \in \mathcal{L}_r} \beta_i y_T^i = 0 \qquad (8.12)$$

$$0 \leqslant \beta_i \leqslant 1$$

where

$$Q = Y J_L K (2\gamma_A I + 2\gamma_I L K)^{-1} J_L^T Y \qquad (8.13)$$

Equation 8.12 is a standard quadratic programming (QP) problem. The optimal solution can be derived utilizing traditional off-the-shelf SVM QP solvers, and we use the SPM:QPC solver[1] in this work.

8.5.3 Classification Error Bound of Related LapSVM

Here we provide a theoretical classification error bound for the proposed related LapSVM via comparison with the established error bound of standard LapSVM.

---

[1] http://sigpromu.org/quadprog/.

An experimental performance evaluation for the related LapSVM is deferred to Section 8.6.

Given a data distribution $\mathcal{D}$ and classifier function class $\mathcal{F}$, the classification error of LapSVM is bounded by the summation of the empirical error, function complexity, and data complexity, as formally stated in the following lemma [175].

Lemma 8.1 *[175]. Fix $\delta \in (0, 1)$ and let $\mathcal{F}$ be a class of functions mapping from an input space $\mathcal{X}$ to $[0, 1]$. Let $\{x_i\}_{i=1}^l$ be drawn independently according to a probability distribution $\mathcal{D}$. Then with a probability of at least $1 - \delta$ over random draws of samples of size l, every $f \in \mathcal{F}$ satisfies*

$$E_{\mathcal{D}}[f(x)] \leq \hat{E}[f(x)] + R_l(\mathcal{F}) + \sqrt{\frac{\ln(2/\delta)}{2l}} \qquad (8.14)$$

where $\hat{E}[f(x)]$ is the empirical error averaged on the l examples and $R_l(\mathcal{F})$ denotes the Rademacher complexity of the function class $\mathcal{F}$.

By utilizing the error bound of SVM [43], $\hat{E}[f(z)] \leq O(\|\xi\|_2^2 \log^2 l)$, we can further bound the error of LapSVM in terms of the slack variable ε_i as follows:

$$E_D[f(x)] \leq O\left(\sum_i \varepsilon_i^2 \log^2 l\right) + R_l(\mathcal{F}) + \sqrt{\frac{\ln(2/\delta)}{2l}}$$

where ε_i is the slack variable for sample x_i in the labeled or related sample set. The proposed related LapSVM reduces the classification error bound over LapSVM via properly reweighting the slack variable for the unconfident/noisy samples. Specifically, consider the case where a sample x_j is selected as a confident sample but labeled incorrectly. For x_j, training the classifier actually minimizes an incorrect slack variable ε_j and maximizes the correct slack variable $\hat{\varepsilon}_j = 1 - \varepsilon_j$, due to its opposite label. $\hat{\varepsilon}_j$ is maximized within the range of $[0, 2]$. Thus, the error bound is increased to

$$E_D[f(z)] \leq O\left(\left(\sum_{i \neq j} \varepsilon_i^2 + \hat{\varepsilon}_j^2\right) \log^2 l\right) + R_l(\mathcal{F}) + \sqrt{\frac{\ln(2/\delta)}{2l}}$$

In contrast, related LapSVM reduces the feasible range of $\hat{\varepsilon}_j$ to $[0, 2 - \rho \cdot C_T^j]$. Consequently, the value of $\hat{\varepsilon}_j$ is decreased, and we have a lower error bound for related LapSVM than for standard LapSVM:

$$E_D[f_{\text{re-LapSVM}}(x)] \leq E_D[f_{\text{LapSVM}}(x)] \qquad (8.15)$$

The above analysis can be generalized to the case where more unlabeled samples are labeled incorrectly. Thus, we can conclude that the related LapSVM reduces error bound via handling the incorrectly labeled samples better.

8.6 Experiments

We conducted extensive experiments to evaluate the effectiveness of the proposed adaptive learning method for celebrity identification. This section is organized as follows: Section 8.6.1 introduces the details of construction of the used database. We demonstrate the experimental settings in details in Section 8.6.2. Section 8.6.3 shows a naive approach of including the video constraint in building the sample affinity graph and demonstrates that video context can improve the performance to a limited degree. Sections 8.6.4 and 8.6.5 show the effectiveness of related samples in both supervised and semisupervised learning scenarios. The average precision is reported on both image and video testing sets. Section 8.6.6 illustrates the performance curve of the proposed method along with learning iterations. We include in the last subsection experiments of related samples on a public database: YouTube Celebrity Database.

8.6.1 Database Construction

Since databases with sufficient image and video samples for celebrity identification are rare, in this work, we construct a database for benchmarking different methods for this task as shown in Figure 8.4. The collection of image and video data is described as follows.

8.6.1.1 Image Data

We selected 30 celebrities who are well known within their fields so that sufficient corresponding video data can be crawled. For each individual, we retrieved about 100 clear images from Google Images using the names of celebrities as query. We manually labeled all the images and marked the locations of eyes. All faces we are then normalized via a standard affine transformation. There was no strict constraint in photography conditions—different poses, facial expressions, and illumination conditions we all allowed. Fifteen images were randomly sampled to form the training image set, while the remaining were used as the testing set. We report the average precision (AP) from five different random training–testing splits. The list of celebrities chosen in the database is given in Table 8.1.

8.6.1.2 Video Data

Querying by the names of the celebrities, a video corpus consisting of about 300 video clips was downloaded from video sharing websites, for example, YouTube. Note that for the experiments, we assume that the videos are unlabeled for the following reasons: (1) the keyword searching results are not reliable, and videos are not necessarily related to the celebrities, and (2) there may also be individuals other than the celebrities of interest in the returned videos.

In this chapter, only the detected face tracks were considered in the iterative adaptive learning process. Thus, based on the video constraint, the label was transferred

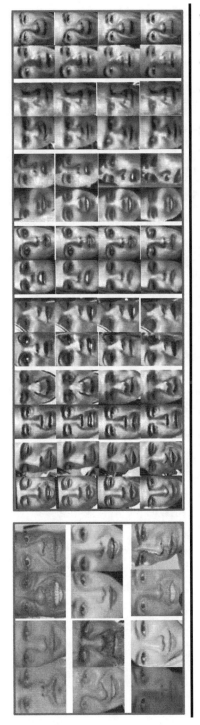

Figure 8.4 Examples in image (left) and video (right) databases. Variations in the video set are more significant than those in the image set.

Table 8.1 Celebrities Included

Occupation	Name	Gender	Video Source
Politician	Barack Obama	M	Speech News report
	Yingjiu Ma	M	
	Al Sharpton	M	
Western actor	Adam Sandler	M	Movies Interviews
	Alexander Skarsgard	M	
	Alan Alda	M	
	Anthony Hopkins	M	
	Alan Rickman	M	
	Alan Thicke	M	
	Amy Poehler	F	
	Alicia Silverstone	F	
Asian actor	Chao Deng	M	Movies Interviews
	Baoqiang Wang	M	
	Zidan Zhen	M	
	Benshan Zhao	M	
	Bingbing Fan	F	
	Wei Tang	F	
	Yuanyuan Gao	F	
Singer	Dehua Liu	M	Music albums Concerts
	Katy Perry	F	
	Wenwei Mo	F	
	Xiaochun Chen	M	
	Yanzi Sun	F	

(Continued)

Table 8.1 Celebrities Included (*Continued*)

Occupation	Name	Gender	Video Source
Host and anchor	Anderson Cooper	M	News report
	Fujian Bi	M	Talk show TV programs
	Lan Yang	F	
	Jing Chai	F	
CEO	Yun Ma	M	Commercial news
	Bill Gates	M	Product launch video
	Steve Jobs	M	

We chose people with different occupations, as listed above. For different occupations, video data were collected from different video sources correspondingly.

from confident frames to uncertain frames within the same track. Besides, by only considering the detected tracks, the volume of frames that needed to be processed could be largely reduced to accelerate the learning process. To obtain reliable face tracks, a robust foreground correspondence tracker [187] was applied for each shot.

Here, video shot segmentations were automatically detected with the accelerating shot boundary detection method [60]. More specifically, the focus region (FR) in each frame was defined, and using a skip interval of 40 frames, the method not only sped up the detection process, but also found more subtle transitions.

After segmenting the video into shots, the tracking process took the results of OKAO face detection[1] as input and generated several face tracks using the tracking algorithm in [187]. The face tracks were then further pruned via fine analysis of faces as follows:

- **Duration**. Short tracks with less than 30 frames were discarded since these tracks are often introduced by false positive detections.
- **Clusters**. K-means clustering was applied to each track, and only those frames closest to clustering centers were chosen as corresponding representative faces.

Consequently, we acquired around 2700 video tracks in total, with nearly 90 tracks per individual.

---

[1] http://www.omron.com/rd/coretech/vision/okao.html.

8.6.1.3 Feature for Face Recognition

We adopted the following three types of state-of-the-art features in face recognition: Gabor, local binary pattern (LBP), and scale-invariant feature transform (SIFT). Details of the feature extraction are listed below:

- **Gabor feature**. The Gabor filter [46] has been widely used for facial feature extraction due to its capability to capture salient visual properties, such as spatial localization, orientation selectivity, and spatial frequency characteristics. In this chapter, we adopted a common setting for extracting Gabor features: wavelet filter bank with five scales and eight orientations, central frequency set at $\sqrt{2}$, and filter window width set at 2π.
- **Local binary pattern feature**. LBP captures the contrast information of the central pixel and its neighbors. The advantage of LBP lies in its robustness to illumination and pose variations. We used a variant of LBP: multiblock LBP [142]. In the feature selection, the image was first segmented into several blocks to keep a certain amount of geometric information. Each face image was divided into 5×4 subregions, and then for each subregion, uniform patterns were extracted and concatenated as bins for a histogram representation.
- **SIFT feature**. A nine-point SIFT feature was used in the experiments. Referring to the work of Everingham et al. [54], a generative model was adopted to locate the nine facial key points in the detected face region, including the left and right corners of each eye, the two nostrils and the tip of the nose, and the left and right corners of the mouth, followed by the 128-dimension SIFT feature [124] extraction process.

The vectors of the above three features were normalized individually by the l_2-norm and concatenated into a single vector for each image/frame.

8.6.2 Experiment Settings

In the following experiments, the initial training image set was constructed by randomly sampling 15 images per person from the labeled image data, and the rest of the images were used for testing. We ran this sampling process five times in each experiment and report the mean precision in this chapter.

We considered two scenarios for experiments: 10-person and 30-person scenarios. In the 10-person scenario, 10 celebrities are selected randomly from the name list in Table 8.1 and corresponding training samples were chosen as above. We performed random selection processes three times and then reported the average precision (AP). In the 30-person scenario, we used the training samples of all celebrities.

For adaptive learning (AL), we followed the procedures in Algorithm 8.1 with the value of parameter S set as 5. The maximal iteration number was set at $N_{iter} = 15$, and the results for AL-based approaches were the accuracy of the final learning iteration. The parameter η in Equation 8.1 was set at 0.7. In related LapSVM, defined in Equation 8.6, γ_I and γ_A were set at 10^{-2} and 1, respectively, and ρ was empirically set at 0.3.

8.6.3 Video Constraint in Graph

LapSVM [132] is a graph-based classifier, and we took a baseline extension to incorporate the video context information into the LapSVM framework.

The general idea was to include the video constraint when constructing the affinity matrix, which defines the similarity among training instances. A naive approach is to set the similarity of frames from the same track to 1. Nevertheless, experiments show that this setting usually results in a degradation in performance. A possible reason could be that the weight among consecutive frames becomes much larger than that of other entries within the weight matrix, which makes the classifier dominated by the constraints on corresponding samples other than labeled instances. Therefore, to ensure the balance of sample weights, the weight was defined as the summation of graphic similarity and video constraint. In detail, the edge between consecutive frames was defined as

$$w_{ij} = \lambda \cdot \exp\{-(x_i - x_j)^2/2\sigma^2\} + (1 - \lambda) \cdot \min\{\zeta \cdot \mu_W, 1\} \qquad (8.16)$$

where μ_W is the mean of matrix **W**.

In Equation 8.16, we confine $\lambda \in [0, 1]$ and $\zeta \in [1, 10]$ empirically. We tune the values for λ with a step of 0.1 and ζ with a step of 1 within their corresponding ranges via cross-validation. Experiments show a small improvement over LapSVM of 1% on average.

This approach is named Lap+V and is taken as the baseline algorithm in the following experiments.

8.6.4 Related Sample in Supervised Learning

In this section, we evaluate the effect of related sample on SVM. γ_I in Equation 8.6 is set at 0 and the classifier is defined only in terms of labeled samples $f(\cdot) = \sum_{i=1}^{l} \alpha_i K(x_i, \cdot)$. We compare the following methods: SVM, ST-SVM (self-training with SVM), AL-SVM (adaptive learning with video constraint), and Re-SVM (related SVM). Similar to Section 8.6.3, the performance was evaluated on both image and video data in 10-person and 30-person scenarios, respectively. Average precisions are reported in Tables 8.2 and 8.3 (the best results in both image set and video set are in bold), under varying numbers of labeled training images.

For traditional self-training, only those frames with high similarity to the initial training samples were selected to enlarge the training set. Thus, the variations in the selection samples were limited. A limited number of labeled samples may decrease the AP due to the high error rate during selection, while more labeled samples usually results in improvement for ST-SVM. However, the difference for either

Table 8.2   Comparison of the Average Precisions (%) of Different SVM-Based Methods in the 10-Person Scenario

		3	5	7	10	12	15
Image	Lap+V	**50.83**	**60**	**68.33**	76.67	82.5	84.17
	SVM	39.17	48.75	58.75	75	75.83	78.75
	ST-SVM	41.67	51.25	61.25	75.42	76.25	80.83
	AL-SVM	43.34	51.25	58.75	77.92	80.84	82.09
	Re-SVM	50.42	54.17	65.42	**82.5**	**82.92**	**84.59**
Video	Lap+V	**53.09**	48.38	49.97	56.62	70.32	73.41
	SVM	30.4	33.09	45.44	62.23	66.76	70.68
	ST-SVM	29.11	34.17	44.21	64.12	66.88	72.93
	AL-SVM	46.32	50.3	45.84	55.73	63.37	75.42
	Re-SVM	49.55	**50.45**	**50.57**	**76.26**	**79.26**	**78.18**

Table 8.3   Comparison of the Average Precisions (%) of Different SVM-Based Methods in the 30-Person Scenario

		3	5	7	10	12	15
Image	Lap+V	**48.87**	**62.83**	**72**	**82**	**84.5**	**86.5**
	SVM	39.33	51.17	61.67	75	79.5	82.17
	ST-SVM	39.33	50.33	60.5	75.33	79	81.67
	AL-SVM	38.34	48.17	57.34	72.17	79	84.5
	Re-SVM	41.5	52.84	63	76.34	82.5	84.34
Video	Lap+V	42.13	46.48	45.93	56.61	62.39	60.82
	SVM	31.97	38.79	41.41	50.88	60.48	64.95
	ST-SVM	32.26	38.08	40.74	51.55	59.67	63.61
	AL-SVM	36.06	38.35	48.95	64.01	69.55	74.02
	Re-SVM	**49.28**	**46.67**	**49.09**	**67.03**	71.47	**75.42**

degradation or improvement is very small: less than 1%. Straightforward adaptive learning (AL+SVM) demonstrates a similar performance, but the ranges for both degradation and improvement are largely increased to around 4%.

Related SVM adjusts the margin for each sample in accordance with its confidence scores, such that we can amplify the positive influence of more confident samples while suppressing the negative influence of less confident samples. Generally, by regarding selected samples as related samples, the classifier is much more robust to selection errors. As shown in Tables 8.2 and 8.3, the improvement of Re-SVM over SVM is around 5% on image data and 12% on video data. In most cases where the number of labeled samples is small (e.g., three or five), the initial classifier is unreliable. Normally, around half of the selected tracks are not correctly labeled by the classifier. Related SVM can significantly degrade the impact of error tracks and provide considerable AP improvement. With sufficient labeled training samples (e.g., 12 or 15), the generalization performance of the classifier is significantly improved. The error rate in selecting tracks is low, and thus correct samples play a dominant role in training. In such a case, the improvement brought by related samples becomes less significant.

Note that there is still a considerable performance gap between related SVM and LapSVM with video constraint (Lap+V) on the image testing dataset: 3% and 6% in 10-person and 30-person cases, respectively. A possible reason lies in the fact that both training and testing samples are static images downloaded from Google. The correlation between video data and image data is low. As a consequence, the right tracks selected in AL will result in minor improvement for testing on images, while the incorrect tracks will degrade the performance to a certain extent. The impact of error tracks is relatively significant compared with the influence of the right tracks. However, on the video dataset, related SVM outperforms LapSVM with margins of 5% and 7% in 10-person and 30-person cases, respectively. Especially, when sufficient labeled samples are fed into the training process—10 or more—the improvement can be up to 20%.

8.6.5 Related Samples in Semisupervised Learning

In this section, we examine the effect of related samples in semisupervised learning and take LapSVM and transductive SVM (TSVM) as the base classifiers for adaptive learning.

When building up the affinity graph in LapSVM, the video constraint in Equation 8.16 is not included. Related LapSVM (Re-LapSVM) is considered another way of incorporating video constraint into the learning process other than Lap+V in Section 8.6.3. The video context information is utilized in the process of promoting tracks into the related set.

We investigate whether further improvement of LapSVM can be brought about by Re-LapSVM over Lap+V. The results are given in Tables 8.4 and 8.5 (the best results in both image set and video set are in bold). We observe similar results in

Table 8.4 Comparison on the Average Precisions (%) of Different LapSVM-based Methods in the 10-Person Scenario

		3	5	7	10	12	15
Image	Lap+V	50.83	60	68.33	76.67	82.5	84.17
	ST-LapSVM	48.75	56.67	66.25	77.5	78.75	82.08
	AL-LapSVM	**50.84**	46.67	73.34	81.25	82.92	87.92
	Re-LapSVM	49.17	**61.67**	**75.84**	**83.34**	**85.42**	**88.34**
Video	Lap+V	53.09	48.38	49.97	56.62	70.32	73.41
	ST-LapSVM	34.83	42.39	48.62	66.34	71.19	75.72
	AL-LapSVM	48.21	16.67	39.27	64.06	63.76	79.83
	Re-LapSVM	**53.46**	**55.28**	**77.01**	**83.13**	**83.28**	**84.36**

comparisons among self-training with LapSVM (ST-LapSVM), straightforward AL (AL-LapSVM), and AL with related samples (Re-LapSVM). Re-LapSVM outperforms both ST-LapSVM and AL-LapSVM. Re-LapSVM demonstrates a better tolerance to selection errors than AL-LapSVM, especially for cases with three, five, and seven labeled samples. More importantly, the comparison between Lap+V and Re-LapSVM demonstrates more insightful results. Re-LapSVM demonstrates

Table 8.5 Comparison of the Average Precisions (%) of Different LapSVM-Based Methods in the 30-Person Scenario

		3	5	7	10	12	15
Image	Lap+V	48.87	62.83	72	82	84.5	86.5
	ST-LapSVM	45.33	60	72.17	82.33	84.67	86.67
	AL-LapSVM	40.17	52.34	64.17	76.67	78.34	84
	Re-LapSVM	**49**	**64.67**	**72.84**	**83.84**	**84.67**	**87.5**
Video	Lap+V	42.13	46.48	45.93	56.61	62.39	60.82
	ST-LapSVM	38.22	49.5	59.18	**64.76**	70.38	69.71
	AL-LapSVM	26.41	41.52	39.75	57.32	62.68	61.75
	Re-LapSVM	**42.29**	**50.66**	**57.16**	63.33	**71.52**	**72.9**

Table 8.6 Comparison of the Average Precisions (%) of Different TSVM-Based Methods in the 10-Person Scenario

		3	5	7	10	12	15
Image	TSVM	41.67	51.67	60.42	75.83	77.08	79.58
	AL-TSVM	39.58	48.75	59.17	74.58	79.17	85
	Re-TSVM	**42.5**	**52.08**	**62.5**	**80**	**78.75**	**85**
Video	TSVM	32.01	38.82	46.91	61.87	68.11	71.94
	AL-TSVM	**39.24**	40.05	43.05	61	73.56	78.66
	Re-TSVM	36.06	**42.66**	**47.39**	**75.03**	**75.99**	**81.92**

a significant advantage over Lap+V. In details, the enhancement on AP is around 4% for the 10-person image case, 1.5% for the 30-person image case, 16% for the 10-person video case, and 8% for the 30-person video case.

In the implementation of TSVM, we optimize TSVM following Collobert et al. [40] with a concave–convex procedure (CCCP). The objective function of TSVM is nonconvex, and CCCP optimizes the problem by solving multiple quadratic programming subproblems. For each QP subproblem, we incorporate related samples in a manner similar to that in Equation 8.6. Since the optimization of TSVM is slow, we only conduct experiments in the 10-person scenario. The results are demonstrated in Table 8.6 (the best results in both image set and video set are in bold).

Clearly, the performance of TSVM is worse than that of LapSVM, especially when labeled samples are limited. However, the comparison between TSVM and LapSVM is out of the scope of this work. Here, our focus is on whether related samples improve the performance of TSVM as well. As shown in Table 8.6, Re-TSVM outperforms both TSVM and AL-TSVM with an improvement of around 3%.

8.6.6 Learning Curves of Adaptive Learning

In this section, we investigate the behaviors of different approaches by investigating the average precision with respect to the iteration number. In this experiment, the labeled set for testing and training was fixed for a fair comparison. The maximal iteration count was set at 15, and accuracy on testing data was reported for each iteration. Since the learning curve was similar for most simulation runs, Figure 8.5 illustrates one run for LapSVM-based adaptive learning.

It is easy to observe that straightforward adaptive learning (naive AL) shows a noisy curve since it is quite sensitive to selection errors. If the correct track is chosen, accuracy will demonstrate an obvious increase, and the performance will drop suddenly if errors occur in the process of selection. Re-LapSVM with $\rho = 0$

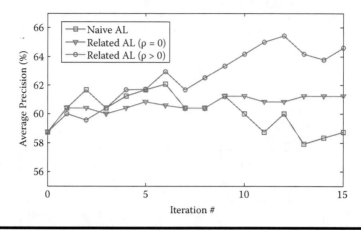

Figure 8.5 **Learning curves of three approaches: naive AL, related AL ($\rho = 0$), and related AL ($\rho > 0$).**

shows a smooth learning curve and converges. Re-LapSVM with $\rho > 0$ shares behaviors similar to those of the two approaches to some extent: the trend of AP is increasing, but with minor turbulence. The parameter ρ in Equation 8.6 is an important factor controlling the relative influence compared with the labeled image samples in the learning process. A larger ρ will render the learning curve closer to straightforward AL, while a smaller ρ pushes the learning curve toward a related AL with $\rho = 0$. An exemplar illustration of simulation results is presented in Figure 8.6. In general, the observed results are consistent with our expectation.

8.6.7 YouTube Celebrity Dataset

We also evaluated the proposed algorithm on a public dataset—YouTube Celebrity Dataset [94]—which contains 1910 sequences of 47 subjects. All the sequences are extracted from video clips downloaded from YouTube by evicting frames that do not contain celebrites of interest. Most of the videos are of low resolution and recorded at high compression rates. The size of frames ranges from 180×240 to 240×320 pixels.

Following methods similar to those described in Section 8.6.1, face tracks were extracted within each video sequence. Only celebrities with more than 30 tracks were included in this experiment, and the final number of identities was 32. Since there is no separate image set for the initial training stage in our approach, we randomly sampled five tracks for each celebrity. All the frames within were then treated as initial labeled samples. This sampling process was repeated five times, and the corresponding averaged results are shown in Table 8.7. The results are similar to those observed on our own dataset, and the improvement of Re-LapSVM over Lap+V is around 4% on average.

Figure 8.6 Examples of iterative improvement. The upper left static images are used for training the initial classifier, and the gray image matrix represents the pool of video tracks with each column standing for a track. In iteration 1, tracks in the blue bounding box were chosen, while in iteration 2, tracks in the orange bounding box were selected. The lowermost row shows examples of testing images, with corresponding confidence scores shown below. A red frame indicates a wrong decision, and a green frame indicates a right decision. With more tracks selected in the training pool, the confidence score on the testing dataset is rising.

Compared with the results of our own dataset, the improvement of related LapSVM is less significant over the baseline algorithms. The reason is that the proposed method targets solving a common problem in real applications; namely, it is difficult to collect many training images to train reliable initial classifiers. When the number of labeled training images is small, the classifiers are not reliable, and errors

Table 8.7 Comparison of the Average Precision (%) of Different LapSVM-Based Methods in the YouTube Celebrity Database

	3	5	7	10	12	15
SVM	45.73	47.3	55.05	57.66	62.25	41.33
ST–SVM	43.35	47.47	55.07	57.39	63.36	63.5
Lap+V	49.97	50.93	60.37	63.83	67.81	68.25
ST-LapSVM	51.3	52.3	61.55	63.92	68.05	68.29
AL-LapSVM	55.93	55.72	62.02	65.33	68.59	68.8
Re-LapSVM	**58**	**58.09**	**65.61**	**65.81**	**69.81**	**68.87**

in selecting the confident video tracks by such weak initial classifiers are inevitable. In this case, the performance of classifiers may degrade severely due to incorporating more and more noisy or incorrect samples. Thus, the improvement brought by related LapSVM is more significant with more noisy tracks selected.

Compared with the proposed dataset, the error rate in selecting confident tracks from the YouTube dataset is much lower. Thus, the performance gain of Re-LapSVM is smaller on the YouTube dataset than on our own dataset. The reasons for lower track selection error on the YouTube dataset are twofold: (1) The YouTube dataset only contains videos, so we train the initial classifier using the video data. Such video-domain classifiers perform more accurately in selecting the confident remaining video tracks than the initial classifiers trained from the image domain in our own dataset; and (2) The face sequences (tracks) for each individual in the YouTube face dataset are usually extracted from only two or three videos, and the correlation/similarity among different sequences from the same video is quite high. However, the dataset built in this work contains tracks from about 10 different videos for each celebrity. Thus, our video dataset is much more diverse and difficult for track selection than the YouTube face dataset. Due to the above reasons, the performance improvement on the YouTube dataset achieved by Re-LapSVM is less significant than that on our dataset.

8.7 Summary

A novel adaptive learning framework was proposed for the celebrity identification problem inspired by the concept of baby learning. The classifier was initially trained on labeled static images and gradually improved by augmenting confident face tracks into the knowledge base. We also proposed a robust classifier that is robust to selection errors by assigning weak adaptive margin for those selected samples. Extensive experiments were conducted in both supervised and semisupervised learning settings for celebrity identification. The results from two databases show that the improvement in accuracy is significant and inspiring. Although in this work we only consider the task of celebrity identification, the proposed method is a general approach and can be easily extended to solve other problems in computer vision, such as object detection, object recognition, and action recognition.

Chapter 9

Audiovisual Information-Based Highlight Extraction

In video cataloguing, a special but important aspect is highlight extraction, which summarizes a video based on user interest or content importance. Generally, a highlight is something (event or detail) that is of major significance or special interest. With these technologies, users can save time by watching only a summarization of concerned and highlighted parts of a program. In addition, among all types of video, highlight extraction is more popular in sports video, for example, basketball and soccer videos.

9.1 Introduction

With the development of multimedia technology, video resources are increasing rapidly, and extracting the highlights from video files is becoming more and more important. Especially, extracting highlights in sports videos has been paid much attention. Thus, in this chapter, we mainly describe a practical application for highlight extraction in the video summary domain. Basketball is one of the most popular sports programs in our life, and extracting highlights from these games is an important issue in video content analysis. For example, subshots indicating action in a sports program can permit viewers to skip the less interesting portions of the game. Li et al. [105] explore the value of such subshots, where viewers were provided with metadata (manually generated) and instant random access for a wide variety of video content.

In recent years, more and more researchers have focused on extracting more attractive clips based on audio classification, video feature extraction, and highlight modeling for action movies and TV shows. Research for highlight extraction in sports videos has especially been paid much attention. The authors of [156] focused on detecting exciting human speeches and baseball hits, but they did not classify the excitement categories. The authors of [201] proposed a new method for highlight extraction in soccer videos by observing the goalmouth. However, this method can only detect the video highlights and cannot detect the audio highlights. The authors of [194] presented an algorithm to parse the soccer programs into two states (play and break) by using video features. However, it can only discriminate the play and break states simply and cannot extract highlights precisely. Zhao et al. [223] proposed a highlight summarization system for football games based on replays, but it cannot be used in basketball games, where the highlights happen frequently and not all the highlights are replayed. In [197], Xiong et al. used a visual object (e.g., baseball catcher) detection algorithm to find the local, semantic objects in video frames and an audio classification algorithm to find the semantic audio objects in the audio track for sports highlight extraction. Compared with the large and simple game ground in baseball, the basketball ground is more comprehensive and the video background changes more frequently. In addition, the authors of [107] used a hidden Markov model (HMM)-based approach to discover the high-level semantic content of the audio context in action movies. Cai et al. [28] also proposed a framework for a two-stage approach to extract highlights in sports video.

Many related approaches have shown a good performance in specific domains. In order to get a better extraction effect especially in basketball games, in this chapter we propose a novel method that uses both audio and visual features to automatically extract highlight segments. In particular, we set the game of basketball as our initial research target. By analyzing the audiovisual features, we classify the video shot into three types of scene: game scene (GS), studio scene (SS), and ad scene (AS). In the extraction model, we can find the highlight clips in GS by detecting the score changes, audio keywords, and audience's exciting cheers.

9.2 Framework Overview

In this chapter, we combine audio and video information to extract the highlights. As shown in Figure 9.1, the proposed method consists of two major stages: (1) unrelated scene removal and (2) highlight information extraction. In stage 1, we first define three types of user attention scene (i.e., GS, SS, AS) in terms of video and audio features. Since the highlights are always in the game scene, we can remove the studio scene and advertisement scene clips. In stage 2, we extract the highlight segments in each GS by using both video and audio features. We use the visual features to detect the change of score bar and then label the subshot in which the score bar changes as part of a highlight segment. Meanwhile, we find the candidate audio highlight clips

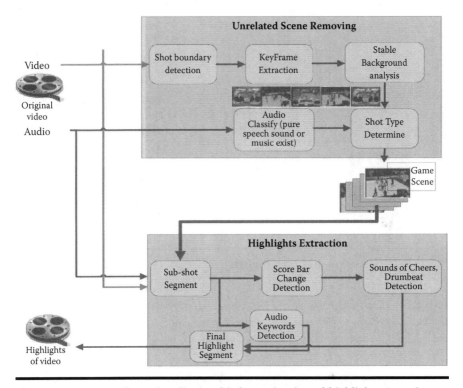

Figure 9.1   System flow of audiovisual information-based highlight extraction.

by detecting exciting sounds (cheering, tantara, drumbeat, etc.) and keywords said by the commentator.

9.3 Unrelated Scene Removal

First, we divide the original video into many shots by the shot boundary detection method. Then we classify all the shots into three types of scenes and remove the unrelated scenes (SS and AS). According to the audio features, we classify each shot into two categories: shot with speech sound and shot with music sound. Meanwhile, by using the visual features, we determine whether a shot has a stable background or not. Figure 9.2 shows the overview of the shot classification process.

9.3.1 Three Types of Scene

It is well known that there exists two phases in a basketball game: the game phase and the suspension phase. As shown in Figure 9.3, in the suspension phase, the commentator's commentary and advertising are usually inserted into the game video.

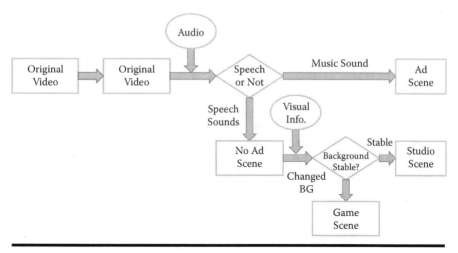

Figure 9.2 Scene classification process.

So, according to the statistics of many basketball game videos, we define three types of user attention scenes in terms of visual and audio features (Figure 9.3). The audiovisual features of each scene are displayed in Table 9.1.

9.3.2 Speech and Nonspeech Discrimination

An essential preliminary step to analyze videos is shot boundary detection. Therefore, the original video is first segmented into several shots with the previously described shot boundary detection methods in Chapter 3. By analyzing many sport videos, we can draw the conclusion that the audio in both GS and SS has speech sound,

Figure 9.3 Three types of scene.

Table 9.1 Audiovisual Features of Scenes

	Game Scene (GS)	*Ad Scene (AS)*	*Studio Scene (SS)*
Audio features	Speech, exciting sound, and keywords	Music and speech	Speech sound
Visual features	Score bar changing	No-score area and unstable background	Stable and fixed background

while there are some music sounds in AS. So in this chapter, we adopt an efficient scheme, proposed by Saad et al. [161], to discriminate speech sounds and music sounds. Because there are speech sounds in every GS, we remove all the shots that don't have speech sounds but have music sounds. After this step, we can remove most of the AS.

9.3.3 Background Differences in SS and GS

The AS is eliminated in Section 9.3.2. We use GS as the final scene type in the second stage, and the next step is to eliminate the SS. We can distinguish SS and GS by analyzing the background because the background in SS is fixed and stable, but diverse in GS. Based on the idea of [133], we design a background difference analysis process as follows.

First, we choose the first frame F_s, the last frame F_e, and the middle frame F_m to represent a shot. Then we use Sobel arithmetic operators to do edge detection in F_s, F_e, and F_m, and we can get the edge maps E_s, E_e, and E_m, which respectively correspond to F_s, F_e, and F_m. The following formula is used to judge the scene type:

$$S_{edge}(E_s, E_m) = \frac{\sum_{(x,y)}(EM_s(x, y) \times EM_m(x, y))}{\sum_{(x,y)EM_s(x,y)}} \qquad (9.1)$$

$$S_{edge}(E_s, E_e) = \frac{\sum_{(x,y)}(EM_s(x, y) \times EM_e(x, y))}{\sum_{(x,y)EM_s(x,y)}} \qquad (9.2)$$

In Equations 9.1 and 9.2, $S_{edge}(E_s, E_m)$ presents the edge match ratio of F_s and F_m, and $S_{edge}(E_s, E_e)$ and $S_{edge}(E_s, E_e)$ are bigger than 0.7, so we can determine that the start frame, middle frame, and end frame are very similar. We can assert that this shot belongs to SS because of its stable and fixed background. By that, SS and AS are removed, and only GS remains as the input of the highlight extraction model.

9.4 Unrelated Scene Removal

Having removed the unrelated scene shot, we extract the highlight clips from the GS. In order to flexibly analyze and extract the highlights, we first divide each shot into subshots in terms of duration. We collect some types of highlights in basketball games (e.g., the most exciting moment, the best slam dunk, and the best pass) to analyze their characteristics. We determine that a subshot is a highlight if

- The score changes
- Keywords (e.g., *hit* and *good*) appear
- The sound of the audience has a sudden increase

The flow diagram is shown in Figure 9.4. At last, we merge the subshot that belongs to a highlight and its neighbor subshots to a whole segment. We call this merged segment a highlight segment (Figure 9.5).

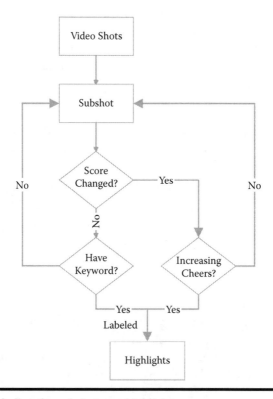

Figure 9.4 Labeling the subshot as a highlight.

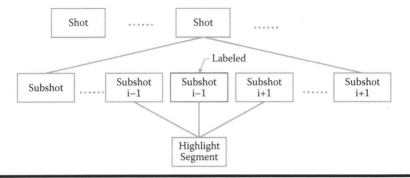

Figure 9.5 Subshot merged to highlight segment.

9.4.1 Subshot Segmenting

To further capture the content changes within a shot, each shot is divided into several subshots. If the length of a shot is more than 10 s and less than 15 s, we divide it into two equal-length subshots; if a shot is more than 15 s, we segment it into three subshots uniformly. For each subshot, we choose the middle frame as its key frame.

9.4.2 Score Bar Change Detection

Analyzing the score information is very important in highlight extraction for sport videos. Many score extraction methods have been proposed in previous works on content-based video analysis [113], and they have been proved to be effective in detecting the score changes. As shown in Figure 9.6, the score and the remaining game time always appear in some fixed positions. In this chapter, we get the score bar area information by using the method described in [113]. Then we detect the change of score bar in each subshot by using the edge match ratio analysis. If the edge match ratio in each subshot is less than the threshold R (we take $R = 0.7$), we can determine that the subshot is a highlight.

Figure 9.6 Marked score bar in the NBA.

9.4.3 Special Audio Words Detection

In a broadcast sports game video like basketball, the exciting audience cheers are closely related to the highlights. In this chapter, we make full use of the prosodic features of audios to detect the increasing cheers. Because exciting speech has both rising pitch and an increasing amount of energy, we use the statistics of pitch and energy extracted from each 0.5 s speech window to judge if there are increasing cheers. Specifically, we use six features: maximum pitch, average pitch, pitch dynamic range, maximum energy, average energy, and energy dynamic range of a given speech window.

However, the audio keyword detection is also an important process in this stage. According to the method proposed by Xu et al. [198], we use the HMM to detect the defined audio keywords. If we detect a keyword, we can determine the subshot belongs to a highlight.

After the above-mentioned audio and visual detection processes, we can recognize the subshot as a highlight. Assuming that a highlight will last for at least 10 s, we merge the labeled subshot and its neighbor subshot to a segment (Figure 9.5). So far, we have finished the extraction work and obtained the highlight segments.

9.5 Experimental Results

In this section, we present the experiment and evaluations of the proposed framework. The testing data were collected from the 2009 NBA playoffs. We chose four basketball game videos, which are listed in Table 9.2. Each video lasts for 15 to 22 min. For each game clip, we took the following parameters to separate the audio and video: video data ratio (211 kbs), video sampling precision (16 bit), audio bit rate (48 kbps), and audio sampling precision (16 bit). Each audio frame length was defined as 40 ms. By analyzing many NBA matches, we selected the keywords as follows: *good goal, one point, two points, three points, hit, score, good shot, well done,* and *dunk.*

In our methods, we tested the ratio of unrelated scene removal, score bar change detection, and audio keyword detection. The performance of unrelated scene removal is listed in Table 9.3. There were many GSs, and few SSs and ASs in each video. We used the text detection method of [133] to get the score area, and then we analyzed the score area's RGB color change to determine whether the score changed

Table 9.2 Tested Basketball Game Videos

Videos	Duration(s)	Dates
Clips 1: Houston Rocket vs. Lakers	985	May 5, 2009
Clips 1: Houston Rocket vs. Lakers	1315	May 9, 2009
Clips 1: Houston Rocket vs. Lakers	1114	May 11, 2009
Clip 4: Cavaliers vs. Orlando Magic	1043	May 25, 2009

Table 9.3   Results of Unrelated Scene Removal

	Clip 1	Clip 2	Clip 3	Clip 4
No. total shot	90	122	97	93
No. detected GS	83	103	87	75
No. true GS	80	99	89	80

Table 9.4   Precision–Recall for Score Change and Keyword Detection

	Score Change	Exciting Audience	Keywords
Precision (%)	97.33	85.71	96.03
Recall (%)	95.67	85.71	93.67

Table 9.5   Accuracy of Audiovisual-Based Framework for Highlight Extraction

	Clip 1	Clip 2	Clip 3	Clip 4
Precision (%)	95.74	91.44	83.71	92.14
Recall (%)	80.36	84.21	79.52	81.17

or not. It presented a perfect precision–recall value. The result is listed in Table 9.4. Moreover, the audio keyword detection stage also reached a high detection accuracy rate of 90%. The overall accuracy detection is detailed in Table 9.5.

It can be seen from Table 9.5 that the precision ratio of our framework is satisfactory except for the recall value, which is a little small. However, the main purpose of our system is to detect and extract highlights from the basketball video as completely as possible, but not leaving out any key highlights. In summary, our experiment results validate the validity of our method; that is, the novel method may effectively eliminate unrelated clips and automatically extract the highlight segments.

9.6 Summary

In this chapter, we proposed a hierarchical method for basketball video highlight clip extraction. In this model, both audio and visual features were used to carry out a comprehensive analysis of sports video from the exclusion of nongame scenes. By flexibly using two stages (unrelated scene removal and highlight shot extraction) and two clues (audio clue and visual clue), the highlight clips were effectively extracted. Compared with methods that only use audio features or visual features, our method achieves both an efficient extraction result and a high accuracy rate.

Chapter 10

Demo System of Automatic Movie or Teleplay Cataloguing

In order to verify the feasibility of the research results, we designed and implemented a simple demo system of automatic movie and teleplay cataloguing (AMTC) based on the approaches introduced in the previous chapters. Using this demo system, we briefly demonstrate and evaluate the effectiveness of related algorithms.

10.1 Introduction

Recently, various video content has become available in digital format, whether directly filmed using digital equipment or transmitted and stored digitally. The creation of video is easier and cheaper than ever before. This trend is even prevalent in movie and teleplay production. According to an official statistic report of YouTube, more than 6 billion h of video is watched each month, and about 100 h of video is uploaded every minute [212]. Thus, how to efficiently summarize, catalogue, and retrieve video data in these large datasets has become a very crucial and challenging task.

In order to provide people with a comprehensive understanding of the whole story in a video, as well as help television broadcasting producers effectively manage and edit video materials (this can improve the reutilization of video material and programs), video cataloguing was introduced several years ago. Traditional video cataloguing mainly focuses on generating a series of visual content for users to browse and understand the whole story of a video efficiently. However, in this book, what we discuss is more specific to the television broadcasting domain. That is, except for traditional ideas, we focus on video cataloguing, video structure parsing, and video summarization and browsing with different semantic contents.

Generally, in the television broadcasting domain or media asset management (MAM) area, video cataloguing, as well as video summarization, is the most important part, and it is widely applied by and in digital content producers and service providers, broadcasting program creation and production, and TV stations at all levels. The original motivation for most of the work in this book was to help television broadcasting producers improve the reutilization of raw video material, as well as more efficiently manage and retrieve movie and teleplay resources. Thus, in this chapter, we show a coarse demo of movie or teleplay summarization and video cataloguing. With the demo system, the video structure is obtained with shot boundary detection, key frame extraction, and scene detection and the basic semantic contents are also extracted by scene recognition, character recognition, and so on. In addition, metadata of the movie or teleplay are also typed in manually. Finally, all of this gained information, including structure information, basic semantic entities, and video metadata content, is defined as cataloguing items to conduct video summarization and video cataloguing. Compared to other similar systems, our system emphasizes the summary and retrieval of movie and teleplay videos with complex backgrounds and rich semantic contents. This demo system is also an important component of the Key Technology and Application of Media Service Collaborative Processing appraisal achievements certified by China's Ministry of Education (No. 019, NF2011). The main interface of the demo system is shown in Figure 10.1.

10.1.1 Application Scenario

News, movie and teleplay, and sports videos are the top three consumption video types in our daily life. Current research in the field of video analysis and processing mostly focuses on these three types of videos. Generally, news and sports videos generally have more obvious and simple production, editing, and organizational rules, and the story or theme is also relatively simpler; thus, there is more domain knowledge to be used. In contrast, movie and teleplay videos express more colorful and comprehensive stories and have more complex visual and semantic features, which makes their analysis more challenging. In fact, with the movie and teleplay cataloguing proposed in this book, we can handle and analyze movie and teleplay videos more efficiently. In addition, the demo system can also be widely used in many fields, such as television production, broadcasting, storage, and retrieval:

1. **Movie and teleplay production and broadcasting system**. The demo system can be embedded in a production and broadcasting system to provide productive and intelligent movie and teleplay production, delivery, and broadcasting.

Figure 10.1   Main interface of automatic movie and teleplay cataloguing demo.

2. **Content-based video indexing, retrieval, and browsing**. The AMTC demo system is also helpful and practical in large-scale movie and teleplay video retrieval, browsing, and management. That is, it can not only offer video owners more convenient and manageable datasets, but also provide end users with more precise video-on-demand services.

10.1.2 Main Functions of the Demo

Video cataloguing is a very important part of multimedia content management. Video cataloguing technology is gradually developing from the manual mode to automatic cataloguing, especially for movie and teleplay videos with plenty of se- mantic information. As described above, we define video cataloguing as a process of structured analytics on videos, including the analysis of video logic and semantic structure.

The AMTC demo system contains three main functions: (1) summary genera- tion, including cataloguing, fragment browsing, and basic configuration; (2) index- ing, including video structure indexing with video shots, video scene-based structure indexing, and indexing based on scene categories, video text, and character names; and (3) automatic cataloguing to catalogue movies and teleplays according to the video structure parsing and basic semantic extraction results. Catalogue files can be generated to describe video structure and semantic content.

10.2 General Design of the Demo

Video cataloguing holds an important position in media asset management, collecting and refining valuable video and audio data for easy retrieval, reuse, and data interaction. In Chapter 1, we discussed that we can usually describe a movie or teleplay from four aspects: video metadata, video structure data, video semantic data, and video story. Video cataloguing in this book is according to these descriptions to reorganize disordered video resources into clear and ordered forms. Actually, we mainly discuss the second and third aspects in this book. The first aspect, video metadata, belongs to the fixed and global description of a movie or teleplay, namely, the metadata. This description information is determined in video recording and editing processes. The fourth aspect of story refers to the scope of human natural language. Although there is related video story research, in the form of a natural language description of video story, a good way to completely realize it has not been achieved.

It is based on cataloguing files for the summarization and retrieval of movies and teleplays in media asset management. More specifically, we first conduct video structure analysis to obtain the video structure data (including series of shots and video scenes), and then we extract the basic semantic content of shots and video scenes as a unit. Finally, together with the metadata extracted from an Internet movie database (including movie title, director, actor table, release date, and film source) for cataloguing and summarization, we generate a cataloguing material library and a corresponding catalogue file that can be used for video material management, browsing, retrieval, and so on.

10.2.1 Overall Diagram of the Demo

The overall diagram of the demo is shown in Figure 10.2. The demo system consists of a video structure layer, intermediate characteristic layer, and basic video semantics layer. According to these three layers, there are three application functions on top of them: video indexing and browsing, video retrieval, and video cataloguing. Among the three layers, the video structure layer mainly carries on the analysis and processing of the underlying data structure and characteristics of the video to acquire the video shot boundary and scene boundary. By this, we can realize the basic operation of video analysis and processing: video shot segmentation and video scene segmentation. The video intermediate characteristic layer is to further extract semantic features based on video structure. The formation of the intermediate layer includes face detection, face tracking, panoramic key frame extraction, representative local feature extraction,

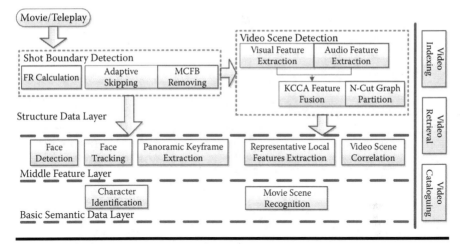

Figure 10.2   Architecture overview of the automatic cataloguing demo.

and video scene association. The basic video semantics layer reorganizes, associates, and recognizes video middle-level feature information to form basic semantic data, such as character recognition, video text, and video scene recognition. By utilizing the characteristics of data and functions in the above three aspects, user-oriented applications, are formed, such as video abstraction and browsing, video indexing, and video cataloguing.

10.2.2 Running Environment

The hardware environment of the system includes the video storage and processing center: Intel® Core™ 2 Quad Q9550 CPU, 2.83 GHz (4-kernel) frequency, 8G memory, GMA 4500M HD graphics card.

The software environment includes:

1. Development environment: Microsoft Visual Studio 2010
2. Senior technical computing language and interactive environment: MATLAB® R2012a
3. Cross-platform build system: CMake
4. Computer Vision Toolbox: OpenCV 2.0

10.2.3 Main Function Modules

As shown in Figure 10.3, the demo system mainly includes three functions: video metadata accessing, video structure parsing, and basic semantic data extraction. There

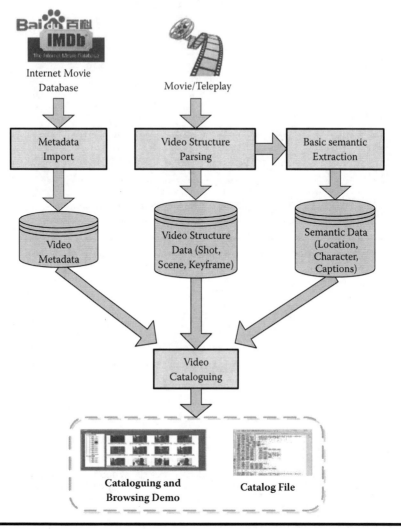

Figure 10.3   System flowchart of the automatic cataloguing demo.

are also six modules: metadata entry, shot boundary detection, video scene detection, video scene recognition, video text extraction, and movie or television character recognition, as shown in Figure 10.4. Finally, with results from these six modules, the catalogue file is formed via video cataloguing for video summarization, browsing, and retrieval. More specifically, we briefly introduce the details of each module as follows:

1. **Metadata entry module**: We collected and stored basic information of the input movie or teleplay, including movie or teleplay name, release date, director,

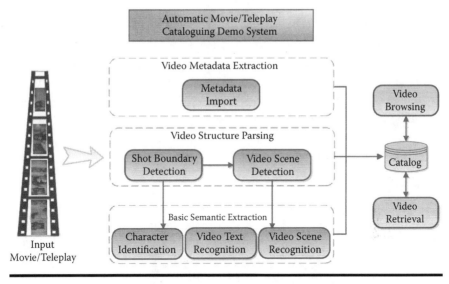

Figure 10.4   Designed framework of the automatic cataloguing demo.

character name table, guidance of public opinion, film source, film length, and copyright. For each movie or teleplay, the metadata always have a unified form; thus, the metadata information were collected one by one manually through an entry dialogue box and eventually written to the catalogue file.

2. **Shot boundary detection module**: We conducted shot boundary detection for the selected movie or teleplay, that is, used the proposed accelerated shot boundary detection method to get all potential shot boundaries, and then parsed the whole video into a sequence of video shots.

3. **Scene detection module**: For all generated shots, by extracting the multi-modality features and establishing similar correlation graphs between shots, we be detected all scene boundaries. That is, video scenes are somehow a cluster of adjacent shots.

4. **Video text recognition module**. We performed video recognition, which implicitly includes video text detection, extraction, and optical character recognition (OCR).

5. **Scene recognition module**. Using a scene recognition method based on the topic model, we classified each video scene into one of the predefined five categories, and the scene category could be used as an entry for summarization and cataloguing.

6. **Movie or teleplay characters recognition module**. We used the actor table information and searched actor pictures for the training set. Then, we formed a face tracker by face detection and tracking in every shot. Finally, we used the multitask joint sparse representation and classification algorithm,

combined with a condition random field model, for face recognition in the face tracker.

By processing with the above-mentioned modules, we could obtain the corresponding metadata, video structure information, and basic semantic information for each movie or teleplay. Next, we used this information to realize the video summarization, browsing, content indexing, and cataloguing.

10.2.4 System Design and Realization

10.2.4.1 Video Abstraction

Generation of video abstract fragments refers to randomly choosing a plurality of video fragments and then splicing them together in accordance with a specific set of rules. The content in a shot always has a unified theme; thus, it is reasonable to choose and orderly splice together multiple shots as video abstract. With shots spliced into a video abstract, the overall content of the original movie or teleplay is well reflected and messy content is avoided. After that, the user can preview and acquire the basic storyline, picture quality, and so on, of the whole movie or teleplay through viewing the video abstract to determine whether he or she is interested in the video on demand and whether he or she wants to buy it.

That is, with video shot segmentation, we can easily form a video summary or abstract by splicing randomly chosen shots. As we mainly deal with movie and teleplay videos, a video summary or abstract is similar to a movie trailer, which combines video fragments at different time points to form a compact clip with content throughout the whole movie or teleplay.

10.2.4.2 Video Indexing and Retrieval

Video indexing, which is similar to a book index, can achieve the purpose of video retrieval. In this chapter, video indexing is divided into three types: directory index (title, time, etc.), structure index (shot, scene, etc.), and semantic content index (character name, video text, scene category, etc.). The demo system mainly provides a structure index and basic semantic content index. Through video shot boundary detection and scene detection, video is partitioned into several shots and scenes in the structure. Therefore, a video structure index is provided at two levels, the shot and scene. In Figure 10.1, the interface on the right side shows a thumbnail of video shots and scenes. These video shots and scenes can also be played in the middle. In addition, by clicking on the "text" and "face" radio buttons, real-time video text detection and face detection can be realized. The system's playback control panel, in addition to the traditional play, pause, and stop buttons and other functions, also provides a fast-forward function on the shot level; that is, by clicking on the fast-forward button, video will fast-forward to the next shot. Besides the structure index, the semantic content index refers to the basic semantic content of a movie or teleplay scene classification and character automatic recognition. The indexed view with different characters and scenes refers to that; by clicking on the main actor names and one of the five kinds of scene categories, the system will display all video footage referring to the selected character names and categories. At the same time, the user can also play the shots and scenes he or she is interested in.

10.2.4.3 Video Cataloguing

Video cataloguing is the main structure of video data and can be realized by using information (including video structure information, basic semantic information, and video metadata), which is generated by the above mentioned modules. In fact, for a input movie or teleplay, cataloguing generates a cataloguing file with the following defined format.

```
CataloguingFile ::= Begincatalogue < Cataloguing Description > < CataloguingContent >
< Cataloguinglog > Endcatalogue
Cataloguing Description ::< Cataloguing Description > < DefinitionUnit >;| < DefinitionUnit >;
DefinitionUnit ::= Definition <_N ame > {< _Meanning >, < _Format >, < _Remarks >};
CataloguingContent ::=< _CataloguingContent :> < Metadata :> < VideoMetadata >
< Stuctureand Semantic Data >


VideoMetadata ::=< VideoMetadata > < MetadataItem > < _Content >;| < MetadataItem >
< _Content >;
MeatadataItem ::=< _Movie/TeleplayName > | < _Release Date > | < _Director >
| < _CharacterList > | < _Guidence > | < _Source > | < _Confidentiality > | < _Copyright >
| < _QualityEvaluation > | < _Length >
Structureand Semantic Data ::=< Structureand Semantic Data :> < _Quality Evaluation >
< VideoScene >
VideoScene ::=< VideoScene > < SceneUnit >;| < SceneUnit >;
SceneUnit ::=< BeginScene > < _UnitNo. > < Scene Description > < SceneContent > < EndScene >
Scene Description ::=< SceneCatagory > < _ Shot >
SceneCatagory ::< _Office > | < _Bedroom > | < _Restaurant > | < _Street > | < _Incar >
SceneContent ::=< SceneContent > < ShotUnit >;| < ShotUnit >
ShotUnit ::= BeginShot < _ShotNo. > < ShotContent > EndShot
ShotContent ::< _FirstFrame > < _EndFrame > < IncludedCharacters > < IncludedCaptions >
IncludedCharacters ::=< IncludedCharacters > < _CharacterName >;| < _CharacterName >;
IncludedCaptions ::=< IncludedCaptions > < _Caption >;| < _Caption >;
Cataloguinglog ::=< _log Description :> {< _Cataloguing Date > < _Cataloguing Status >};
```

In the above definitions, items beginning with "_" are end items, and the remaining are nonend items. The catalogue description defines all cataloguing entries in a shot or video scene, including metadata, video structure, and basic semantic data.

The catalog content gives the specific content of all the video shots, as well as the specific catalog information corresponding to each catalog entry.

Extraction of video structure and basic semantic data is required for a large number of calculations and operations; thus, most of these processes are implemented offline and then stored as structure and basic semantic data in the catalogue database for video cataloguing, searching, and so on. The concrete operating processes are to (1) select one movie or teleplay in the tree structure on the left side of the demo system (Figure 10.1); (2) click on the "cataloguing" button, the system pop-up metadata entry dialogue box; (3) choose or enter the metadata content manually, clicking "OK" to continue cataloguing; (4) wait for the system to generate a catalogue file named "movie/teleplay name.cata," at the same time direct into the catalogue browser dialog; and (5) in the catalogue browser dialog box, click different tab items to achieve different levels of preview.

References

1. J. Mats, O. Chum, M. Urban, and T. Pajdla. Robust wide-baseline stereo from maximally stable extremal regions. *Image and Vision Computing*, 22(10): 761–767, 2002.

2. N. Dalal and B. Triggs. Histograms of oriented gradients form human detection. *Image and Vision Computing*. In *Proceedings of the IEEE Computer Society Conference on Computer Vision and Pattern Recognition*, pp. 886–893, 2005.

3. A. Soffer. Image categorization using texture features. In *Proceedings of the 2011 International Conference on Document Analysis and Recognition*, p. 233C237, 1997.

4. D. Adjeroh, M. C. Lee, N. Banda, and U. Kandaswamy. Adaptive edge-oriented shot boundary detection. *EURASIP Journal on Image and Video Processing*, 2009:1–14, 2009.

5. A. Agarwal and B. Triggs. Hyperfeatures multilevel local coding for visual recognition. In *Proceedings of European Conference on Computer Vision*, pp. 30–43, 2006.

6. A. Akutsu, Y. Tonomura, H. Hashimoto, and Y. Ohba. Video indexing using motion vectors. *Image and Vision Computing*. In *Proceedings of SPIE Visual Communications and Image Processing*, pp. 1522–1530, 1992.

7. A. Aner and J. R. Kender. A unified memory-based approach to cut, dissolve, key frame and scene analysis. In *Proceedings of 2001 International Conference on Image Processing*, pp. 370–373, 2001.

8. D. Anguelov, K. C. Lee, S. B. Göktürk, and B. Sumengen. Contextual identity recognition in personal photo albums. In *Proceedings of IEEE Computer Society Conference on Computer Vision and Pattern Recognition*, pp. 1–7, 2007.

9. O. Arandjelovic and A. Zisserman. Automatic face recognition for film character retrieval in feature-length films. In *Proceedings of the 2005 IEEE Computer Society Conference on Computer Vision and Pattern Recognition*, pp. 860–867, 2005.

10. O. Arandjelovic and A. Zisserman. Automatic face recognition for film character retrieval in feature-length films. In *IEEE Computer Society Conference on Computer Vision and Pattern Recognition*, pp. 860–867, 2005.

11. B. Riberiro and R. Baeza-Yates. *Modern Information Retrieval*. Addison-Wesley/ACM Press, Boston, Massachusetts 1999.

12. L. Ballan, M. Bertini, A. Del Bimbo, and G. Serra. Semantic annotation of soccer videos by visual instance clustering and spatial/temporal reasoning in ontologies. *Multimedia Tools Applications*, 48(2):313–337, 2010.

13. M. Bauml, M. Tapaswi, and R. Stiefelhagen. Semi-supervised learning with constraints for person identification in multimedia data. In *Proceedings of the 2013 IEEE Conference on Computer Vision and Pattern Recognition*, pp. 3602–3609, 2013.

14. M. Bauml, M. Tapaswi, and R. Stiefelhagen. Semi-supervised learning with constraints for person identification in multimedia data. In *IEEE Computer Society Conference on Computer Vision and Pattern Recognition*, pp. 3602–3609, 2013.

139

15. M. Belkin, P. Niyogi, and V. Sindhwani. Manifold regularization: A geometric framework for learning from labeled and unlabeled examples. *Journal of Machine Learning Research*, 7:2399–2434, 2006.

16. Y. Bengio and Y. LeCun. Scaling learning algorithms towards AI. In *Large Scale Kernel Machines*, ed. L. Bottou, O. Chapelle, D. DeCoste, and J. Weston. MIT Press, Cambridge, Massachusetts 2007.

17. A. Berg, J. Edwards, M. Maire, R. White, Y. Teh, E. Learned-Miller, and D. Forsyth. Names and faces in the news. In *IEEE Computer Society Conference on Computer Vision and Pattern Recognition*, pp. II-848–II-854, 2004.

18. A. C. Berg, J. Edwards, M. Maire, R. White, Y. Teh, E. Learned-Miller, and D. Forsyth. Names and faces in the news. In *Proceedings of the 2004 IEEE Computer Society Conference on Computer Vision and Pattern Recognition*, pp. II-848–II-852, 2004.

19. M. Bertini, A. Del Bimbo, and W. Nunziati. Automatic detection of player's identity in soccer videos using faces and text cues. In *Proceedings of the 14th Annual ACM International Conference on Multimedia (MULTIMEDIA '06)*, pp. 663–666, 2006.

20. D. Bita, R. H. Mahmoud, and K. A. Mohammad. A novel fade detection algorithm on h.264/avc compressed domain. *Advances in Image and Video Technology*, pp. 1159–1167, 2006.

21. D. Blei, A. Ng, and M. Jordan. Latent dirichlet allocation. *Journal of Machine Learning Research*, 3:993–1022, 2003.

22. A. Blum and T. Mitchell. Combining labeled and unlabeled data with co-training. In *Proceedings of the Eleventh Annual Conference on Computational Learning Theory (COLT '98)*, pp. 92–100, New York, 1998.

23. P. Bojanowski, F. Bach, I. Laptev, J. Ponce, C. Schmid, and J. Sivic. Finding actors and actions in movies. In *Proceedings of the 2013 IEEE International Conference on Computer Vision*, pp. 2280–2287, 2013.

24. A. Bosch, A. Zisserman, and X. Mu Aoz. Scene classification via plsa. In *Proceedings of 2006 International Conference on Computer Vision*, pp. 517–530, 2006.

25. L. Bourdev and J. Malik. Poselets: Body-part detectors trained using 3D pose annotations. In *Proceedings of 2012 International Conference on Computer Vision*, pp. 1365–1372, 2009.

26. X. U. Cabedo and S. K. Bhattacharjee. Shot detection tools in digital video. In *Proceedings of Non-liner Model Based Image Analysis*, pp. 121–126, Glasgow, July 1998.

27. M. Cai, J. Song, and M. R. Lyu. Color-based clustering for text detection and extraction in image. In *Proceedings of the 2002 International Conference on Image Processing*, pp. 117–120, 2002.

28. S. Cai, S. Jiang, and Q. Huang. A two-stage approach to highlight extraction in sports video by using adaboost and multi-modal. In *Proceedings of the 9th Pacific Rim Conference on Multimedia*, pp. 867–870, 2008.

29. J. Cao and A. Cai. Algorithm for shot boundary detection based on support vector machine in compressed domain. *Acta Electronica Sinica*, 36(1):201–208, 2008.

30. V. T. Chasanis, A. C. Likas, and N. P. Galatsanos. Scene detection in videos using shot clustering and sequence alignment. *IEEE Transactions on Multimedia*, 11(1):89–100, 2009.

31. C.-Y. Chen and K. Grauman. Watching unlabeled video helps learn new human actions from very few labeled snapshots. In *IEEE Computer Society Conference on Computer Vision and Pattern Recognition*, pp. 572–579, 2013.

32. C.-Y. Chen, J.-C. Wang, J.-F. Wang, and C.-P. Chen. An efficient news video browsing system for wireless network application. In *Proceedings of International Conference on Wireless Networks, Communications and Mobile Computing (WiCOM)*, pp. 1377–1381, 2005.

33. M. Y. Chen and A. Hauptmann. Searching for a specific person in broadcast news video. In *Proceedings of the IEEE International Conference on Acoustics, Speech, and Signal Processing,* pp. 1036–1039, 2004.

34. X. Chen, J. Yang, J. Zhang, and A. Waibel. Automatic detection and recognition of signs from natural scenes. *IEEE Transactions on Image Processing,* pp. 87–99, 2004.

35. L. H. Chena, Y. C. Lai, and H. Y. M. Liao. Movie scene segmentation using background information. *Pattern Recognition,* 41(3):1056–1065, 2008.

36. N. Cherniavsky, I. Laptev, J. Sivic, and A. Zisserman. Semi-supervised learning of facial attributes in video. In *European Conference on Computer Vision Workshop,* pp. 43–56, 2010.

37. S.-T. Chiu, G.-S. Lin, and M.-K. Chang. An effective shot boundary detection algorithm for movies and sports. In *Proceedings of the 2008 3rd International Conference on Innovative Computer Information and Control,* pp. 173–176, Dalian, China, June 2008.

38. J. Choi, M. R. Ali, F. Larry, and S. Davis. Adding unlabeled samples to categories by learned attributes. In *IEEE Computer Society Conference on Computer Vision and Pattern Recognition,* pp. 875–882, 2013.

39. A. Coates, H. Lee, and A. Y. Ng. An analysis of single-layer networks in unsupervised feature learning. In *Proceedings of the 14th International Conference on Artificial Intelligence and Statistics,* pp. 215–223, 2011.

40. R. Collobert, F. Sinz, J. Weston, and L. Bottou. Large scale transductive SVMS. *Journal of Machine Learning Research,* 7:1687–1712, 2006.

41. C. Cotsaces, M. A. Gavrielides, and I. Pitas. A survey of recent work in video shot boundary detections. In *Proceedings of 2005 Workshop on Audio-Visual Content and Information Visualization in Digital Libraries (AVIVDiLib '05),* pp. 4–6, June 2005.

42. T. M. Cover and J. A. Thomas. *Elements of Information Theory.* John Wiley & Sons, New York, 1991.

43. N. Cristianini and J. Shawe-Taylor. *An Introduction to Support Vector Machines and Other Kernel-Based Learning Methods.* Cambridge University Press, Cambridge, 2000.

44. A. D. Bimbo. *Visual Information Retrieval.* Morgan Kaufmann, San Francisco, 1999.

45. T. Danisman and A. Alpkocak. Dokuz Eyl'ul university video shot boundary detection at TRECVID 2006. In *Proceedings of the TREC Video Retrieval Evaluation (TRECVID),* pp. 1–6, 2006.

46. J. G. Daugman. Uncertainty relation for resolution in space, spatial-frequency, and orientation optimized by two-dimensional visual cortical filters. *Journal of the Optical Society of America A—Optics Image Science and Vision,* 2(7):1160–1169, 1985.

47. K. G. Derpanis. The Harris corner detector. *New York,* October 2004.

48. N. Dimitrova, H.-J. Zhang, B. Shahraray, I. Sezan, T. Huang, and A. Zakhor. Applications of video-content analysis and retrieval. *Journal of IEEE Multimedia,* 9(4):42–55, 2002.

49. A. Divakaran, R. Radhakrishnan, and K. A. Peker. Motion activity-based extraction of key-frames from video shots. In *Proceedings of the IEEE International Conference on Image Processing,* pp. 932–935, 2002.

50. C. Engels, K. Deschacht, J. H. Becker, T. Tuytelaars, S. Moens, and L. V. Gool. Actions in context. In *Proceedings of 2010 British Machine Vision Conference,* pp. 115.1–115.11, 2010.

51. C. Engels, K. Deschacht, J. H. Becker, T. Tuytelaars, S. Moens, and L. V. Gool. Automatic annotation of unique locations from video and text. In *Proceedings of the British Machine Vision Conference,* pp. 350–353, 2010.

52. B. Epshtein, E. Ofek, and Y. Wexler. Detecting text in natural scenes with stroke width transform. In *Proceedings of the 2010 IEEE Conference on Computer Vision and Pattern Recognition,* pp. 2963–2970, 2010.

53. D. Erhan, Y. Bengio, A. Courville, P.-A. Manzagol, and P. Vincent. Why does unsupervised pre-training help deep learning? *Journal of Machine Learning Research*, 11(2010):625–660, 2010.

54. M. Everingham, J. Sivic, and A. Zisserman. "Hello! my name is . . . buffy"—automatic naming of characters in TV video. In *Proceedings of the 17th British Machine Vision Conference*, pp. 889–908, 2006.

55. A. Fathi, M. F. Balcan, X. Ren, and J. M. Rehg. Combining self training and active learning for video segmentation. In *Proceedings of the British Machine Vision Conference*, pp. 78.1–78.11, 2011.

56. B. Fauvet, P. Bouthemy, P. Gros, and F. Spindler. A geometrical key-frame selection method exploiting dominant motion estimation in video. In *Proceedings of the CVIR Conference*, pp. 419–427, 2004.

57. Fei-Fei Li and P. Perona. A Bayesian hierarchical model for learning natural scene categories. In *Proceedings of 2005 IEEE Computer Society Conference on Computer Vision and Pattern Recognition*, pp. 524–531, 2005.

58. A. M. Ferman and A. M. Tekalp. Two-stage hierarchical video summary extraction using the GoF color descriptor. *IEEE Transactions on Multimedia*, 5(2):244–256, 2003.

59. M. Flickner, H. Sawhney, W. Biblack, J. Ashley, Q. Huang, B. Dom, M. Gorkani, J. Hafner, D. Lee, D. Pekovic, D. Steele, and P. Yanker. Query by image and video content: The OBIC system. *IEEE Computer*, 28(9):23–32, 1995.

60. G. Gao and H. Ma. Accelerating shot boundary detection by reducing spatial and temporal redundant information. In *Proceedings of the IEEE International Conference on Multimedia and Expo*, pp. 1–6, 2011.

61. J. Gao and J. Yang. An adaptive algorithm for text detection from natural scenes. In *Proceedings of the 2001 IEEE Computer Society Conference on Computer Vision and Pattern Recognition*, pp. 84–89, 2001.

62. B. Ghanem, T. Z. Zhang, and N. Ahuja. Robust video registration applied to field-sports video analysis. In *Proceedings of the 2012 IEEE International Conference on Acoustics, Speech and Signal Processing*, pp. 1–4, 2012.

63. J. Gllavata, R. Ewerth, and B. Freisleben. Text detection in images based on unsupervised classification of high-frequency wavelet coefficients. In *Proceedings of the 2004 International Conference on Pattern Recognition*, pp. 425–428, 2004.

64. S. Goldman and Y. Zhou. Enhancing supervised learning with unlabeled data. In *International Conference on Machine Learning*, pp. 327–334, 2000.

65. H. Grabner and H. Bischof. Online boosting and vision. In *IEEE Computer Society Conference on Computer Vision and Pattern Recognition*, pp. 260–267, 2006.

66. H. Grabner, C. Leistner, and H. Bischof. Semi-supervised on-line boosting for robust tracking. In *Proceedings of the 10th European Conference on Computer Vision: Part I, ECCV '08*, pp. 234–247, 2008.

67. L. Graham, Y. Chen, T. Kalyan, J. H. N. Tand, and M. Li. Comparison of some thresholding algorithms for text/background segmentation in difficult document images. In *Proceedings of the 2003 International Conference on Document Analysis and Recognition*, pp. 859–865, 2003.

68. M. R. Greene and A. Oliva. Recognition of natural scenes from global properties: Seeing the forest without representing the trees. *Cognitive Psychology*, 58(2):137–176, 2009.

69. S. Grossberg. Adaptive resonance theory: How a brain learns to consciously attend, learn, and recognize a changing world. *Neural Networks*, 37:1–47, 2013.

70. M. Guillaumin, T. Mensink, J. Verbeek, and C. Schmid. Face recognition from caption-based supervision. *International Journal of Computer Vision*, 96(1):64–82, 2012.

71. L. Gupta, V. Pathangay, A. Dyana, and S. Das. Indoor versus outdoor scene classification using probabilistic neural network. *EURASIP Journal on Applied Signal Processing*, 1:1–8, 2007.

72. R. Hammoud and R. Mohr. A probabilistic framework of selecting effective key frames for video browsing and indexing. In *Proceedings of the International Workshop on Real-Time Image Sequence Analysis*, pp. 79–88, 2000.

73. B. Han. Research on content-based video structure analysis. PhD thesis, Beijing Jiaotong University, 2006.

74. B. Han. Shot boundary detection based on soft computation. PhD thesis, Xidian University, 2006.

75. S. H. Han, K. J. Yoon, and I. S. Kweon. A new technique for shot detection and key frames selection in histogram space. In *12th Workshop on Image Proceeding and Image Understanding*, pp. 475–479, January 2000.

76. A. Hanjalic. Shot-boundary detection: Unraveled and resolved? *IEEE Transactions on Circuits and Systems for Video Technology*, 12(2):90–105, 2002.

77. D. R. Hardoon, S. Szedmak, and J. Shawe-Taylor. Canonical correlation analysis: An overview with application to learning methods. *Neural Computation*, 16(12):2639–2664, 2004.

78. D. C. He and L. Wang. Texture classification using texture spectrum. *Pattern Recognition*, 23(8):905–910, 1990.

79. D. C. He and L. Wang. Texture unit, texture spectrum, and texture analysis. *IEEE Transactions on Geoscience and Remote Sensing*, 28(4):509–512, 1990.

80. W. J. Henga and K. N. Ngan. An object-based shot boundary detection using edge tracing and tracking. *Visual Communication and Image Representation*, 12(3):217–239, 2001.

81. M. Héritier, S. Foucher, and L. Gagnon. Key-places detection and clustering in movies using latent aspects. In *Proceedings of 2007 International Conference on Image Processing*, pp. 225–228, 2007.

82. G. E. Hinton. Training products of experts by minimizing contrastive divergence. *Journal of Neural Computation*, 14(8):1771–1800, 2002.

83. C.-R. Huang, H.-P. Lee, and C.-S. Chen. Shot change detection via local keypoint matching. *IEEE Transactions on Multimedia*, 10(6):1097–1108, 2008.

84. J. Huang, Z. Liu, and Y. Wang. Joint scene classification and segmentation based on hidden Markov model. *IEEE Transactions on Multimedia*, 7(3):538–550, 2005.

85. X. Huang, H. Ma, C. X. Ling, and G. Gao. Detecting both superimposed and scene text with multiple languages and multiple alignments in video. *Multimedia Tools and Applications*, 70(3):1703–1727, 2014.

86. X. Huang, H. Ma, and H. Yuan. A hidden Markov model approach to parsing MTV video shot. In *Proceedings of the 2008 Congress on Image and Signal Processing*, vol. 2, pp. 276–280, May 2008.

87. X. Huang, H. Ma, and H. Zhang. A new video text extraction approach. In *Proceedings of 2009 IEEE International Conference on Multimedia and Expo*, pp. 650–653, 2009.

88. A. Hyvarinen and E. Oja. Independent component analysis: Algorithms and applications. *Neural Networks*, 13(4–5):411–430, 2000.

89. K. Ni, A. Kannan, A. Criminisi, and J. Win. Epitomic location recognition. *IEEE Transactions on Pattern Analysis and Machine Intelligence*, 31(12): 2158–2167, 2009.

90. S. B. Jun, K. Yoon, and H. Y. Lee. Dissolve transition detection algorithm using spatio-temporal distribution of MPEG macro-block types. In *Proceedings of ACM International Conference on Multimedia*, pp. 391–294, 2000.

91. A. Kanak, E. Erzin, Y. Yemez, and A. M. Tekalp. Joint audio-video processing for biometric speaker identification. In *Proceedings of International Conference on Multimedia and Expo*, pp. 561–564, 2003.

92. C. Kim and J. N. Hwang. Object-based video abstraction for video surveillance systems. *IEEE Transactions on Circuits and Systems for Video Technology*, 12:1128–1138, 2002.

93. K. I. Kim, K. Jung, and J. H. Kim. Texture-based approach for text detection in images using support vector machines and continuously adaptive mean shift algorithm. *IEEE Transactions on Pattern Analysis and Machine Intelligence*, 25(12):1631–1639, 2003.

94. M. Kim, S. Kumar, V. Pavlovic, and H. Rowley. Face tracking and recognition with visual constraints in real-world videos. In *IEEE Conference on Computer Vision and Pattern Recognition*, pp. 1–8, 2008.

95. S. M. Kim, J. W. Byun, and C. S. Won. A scene change detection in h.264/avc compression domain. In *Proceedings of 2005 Pacific Rim Conference on Multimedia*, pp. 1072–1082, 2005.

96. K. Kita and T. Wakahara. Binarization of color characters in scene images using k-means clustering and support vector machines. In *Proceedings of 2010 International Conference on Pattern Recognition*, pp. 3183–3186, 2010.

97. I. Koprinska and S. Carrato. Temporal video segmentation: A survey. *Signal Processing: Image Communication*, 16(5):477–500, 2001.

98. D. Kuettel, M. Guillaumin, and V. Ferrari. Segmentation propagation in imagenet. In *European Conference on Computer Vision*, pp. 459–473, 2012.

99. S. Kwon and S. Narayanan. Unsupervised speaker indexing using generic models. *IEEE Transactions on Speech and Audio Processing*, 13(5):1004–1013, 2005.

100. M. Kyperountas, C. Kotropoulos, and I. Pitas. Enhanced eigen-audioframes for audiovisual scene change detection. *IEEE Transactions on Multimedia*, 9(4):785–797, 2007.

101. G. C: Lalor and C. Zhang. Multivariate outlier detection and remediation in geochemical databases. *Science of the Total Enviroment*, 281(1):99–109, 2001.

102. S. Lazebnik, C. Schmid, and J. Ponce. Beyond bags of features: Spatial pyramid matching for recognizing natural scene categories. In *Proceedings of 2006 IEEE Computer Society Conference on Computer Vision and Pattern Recognition*, pp. 2169–2178, 2006.

103. H. Lee, C. Ekanadham, and A. Y. Ng. Sparse deep belief net model for visual area v2. In *Proceedings of Conference on Neural Information Processing Systems*, pp. 873–880, 2008.

104. B. Lehane, N. E. O'Connor, H. Lee, and A. F. Smeaton. Indexing of fictional video content for event detection and summarisation. *Journal on Image and Video Processing*, 2007(2):1–15, 2007.

105. F. C. Li, A. Gupta, E. Sanocki, L. He, and Y. Rui. Browsing digital video. In *Proceedings of the SIGHI Conference on Human Factors in Computing Systems*, pp. 169–176, 2000.

106. H. Li, D. Doermann, and O. Kia. Automatic text detection and tracking in digital video. *IEEE Transactions on Image Processing*, 9(1):147–156, 2000.

107. Q. Li, H. Ma, and K. Zheng. A flexible framework for audio semantic content detection. In *Proceedings of the 9th Pacific Rim Conference on Multimedia*, pp. 915–918, 2008.

108. X. Li and Y. Guo. Adaptive active learning for image classification. *IEEE Transactions on Pattern Analysis and Machine Intelligence*, pp. 859–866, 2013.

109. X. Li, W. Wang, Q. Huang, W. Gao, and L. Qing. A hybrid text segmentaion approach. In *Proceedings of 2009 IEEE International Conference on Multimedia and Expo*, pp. 510–513, 2009.

110. Y. Li, S. S. Narayanan, and C. C. Jay-Kuo. Adaptive speaker identification with audiovisual cues for movie content analysis. *Pattern Recognition Letters*, 25(7):777–791, 2004.

111. Y. N. Li, Z. M. Lu, and X. M. Niu. Fast video shot boundary detection framework employing pre-processing techniques. *IET Image Processing*, 3(3):121–134, 2009.

112. R. Lienhart, S. Pfeiffer, and W. Effelsberg. Scene determination based on video and audio features. In *Proceedings of 1999 IEEE International Conference on Multimedia Computing and Systems*, pp. 685–690, 1999.

113. R. Lienhart and A. Wernicke. Localizing and segmenting text in images and videos. *IEEE Transactions on Circuits and Systems for Video Technology*, 12(4):256–268, 2002.

114. T. Lin, H. J. Zhang, and Q. Y. Shi. Video content representation for shot retrieval and scene extraction. *International Journal of Image and Graphics*, 1(3):507–526, 2001.

115. X. Ling, O. Yuanxin, L. Huan, and X. Zhang. A method for fast shot boundary detection based on SVM. In *Proceedings of the 2008 Congress on Image and Signal Processing*, vol. 2, pp. 445–449, Sanya, China, May 2008.

116. A. Liu and Z. X. Yang. Watching, thinking, reacting: A human-centered framework for movie content analysis. *International Journal of Digital Content Technology and Its Applications (JDCTA)*, 4(5):23–37, 2010.

117. C. Liu, C. Wang, and R. Dai. Text detection in images based on unsupervised classification of edge-based features. In *Proceedings of the Eighth International Conference on Document Analysis and Recognition*, pp. 610–614, 2005.

118. H. Y. Liu and H. Zhang. A content-based broadcasted sports video retrieval system using multiple modalities: Sportbr. In *Proceedings of the Fifth International Conference on Computer and Information Technology*, pp. 652–656, 2005.

119. H. Y. Liu and D. R. Zhou. A content-based news video browsing and retrieval system. In *Proceedings of 3rd International Symposium on Image and Signal Processing and Analysis*, pp. 793–798, 2003.

120. J. Liu and M. Shah. Scene modeling using co-clustering. In *Proceedings of the 11th International Conference on Computer Vision*, pp. 1–7, 2007.

121. T. Liu, X. Zhang, J. Feng, and K. Lo. Shot reconstruction degree: A novel criterion for keyframe selection. *Pattern Recognition Letter*, 25(12):1451–1457, 2004.

122. Y. Liu, W. Wang, W. Gao, and W. Zeng. A novel compressed domain shot segmentation algorithm on h.264/avc. In *Proceedings of 2004 International Conference on Image Processing*, pp. 2235–2238, 2004.

123. Z. Liu and Y. Wang. Major cast detection in video using both speaker and face information. *IEEE Transactions on Multimedia*, 9(1):89–101, 2007.

124. D. G. Lowe. Object recognition from local scale-invariant features. In *Proceedings of the 1999 International Conference on Computer Vision*, pp. 1150–1157, 1999.

125. D. G. Lowe. Distinctive image features from scale-invariant keypoints. *International Journal of Computer Vision*, 60(2):91–110, 2004.

126. J. B. MacQueen. Some methods for classification and analysis of multivariate observations. In *Proceedings of 5th Berkeley Symposium on Mathematical Statistics and Probability*, pp. 281–297, 1967.

127. M. K. Mandal, F. Idris, and S. Panchanathan. Algorithm for shot boundary detection based on support vector machine in compressed domain. *Image and Vision Computing*, 17(7):513–529, 1999.

128. V. Y. Mariano and R. Kasturi. Locating uniform-colored text in video frames. In *Proceedings of the 15th International Conference on Pattern Recognition*, pp. 539–542, 2000.

129. M. Marszalek, I. Laptev, and C. Schmid. Actions in context. In *Proceedings of 2008 IEEE Conference on Computer Vision and Pattern Recognition*, pp. 2929–2936, 2009.

130. J. Mas and G. Fernandez. Video shot boundary detection based on color histogram. In *Proceedings of the TREC Video Retrieval Evaluation Conference (TRECVID 2003)*, pp. 2.1–2.11, November 2003.

131. D. McClosky, E. Charniak, and M. Johnson. Effective self-training for parsing. In *2006 Conference of the North American Chapter of the Association for Computational Linguistics: Human Language Technologies*, pp. 152–159, 2006.

132. S. Melacci and M. Belkin. Laplacian support vector machines trained in the primal. *Journal of Machine Learning Research*, 12:1149–1184, 2011.

133. C. Mi, Y. Liu, and X. Xue. Video texts tracking and segmentation based on multiple frames. *Journal of Computer Research and Development*, 43(9):1523–1529, 2006.

134. G. Miao, Q. Huang, S. Jiang, and W. Gao. Detecting text in natural scenes with stroke width transform. In *Proceedings of the 2008 IEEE International Conference on Multimedia and Expo*, pp. 569–572, 2008.

135. K. Mikolajczyk and C. Schmid. Scale and affine invariant interest point detectors. *International Journal of Computer Vision*, 60(1):63–86, 2004.

136. J. M. Zhang, L. R. Yang, and L. M. Wang. Facial expression recognition based on hybrid features fused by CCA. *Computer Applications*, 28(3), 2008.

137. H. Q. Minh, L. Bazzani, and V. Murino. A unifying framework for vector-valued manifold regularization and multi-view learning. In *Proceedings of the 30th International Conference on Machine Learning*, vol. 28, pp. 100–108, May 2013.

138. A. Mostefaoui, H. Kosch, and L. Brunie. Semantic based prefetching in news-on-demand video servers. *Multimedia Tools and Applications*, 18(2):159–179, 2002.

139. M. R. Lyu, J. Song, and M. Cai. A comprehensive method for multilingual video text detection, localization, and extraction. *IEEE Transactions on Circuits and Systems for Video Technology*, 15(2):243–255, 2005.

140. C. H. Nge, T. C. Pong, and H. J. Zhang. On clustering and retrieval of video shots. In *Proceedings of 2001 ACM International Conference on Multimedia*, pp. 51–60, 2001.

141. T. Ojala, M. Pietikainen, and T. Maenpaa. Multiresolution gray-scale and rotation invariant texture classification with local binary patterns. *IEEE Transactions on Pattern Analysis and Machine Intelligence*, 24(7):971–987, 2002.

142. T. Ojala, M. Pietikinen, and T. Menp. Multiresolution grayscale and rotation invariant texture classification with local binary patterns. *IEEE Transcations on Pattern Analysis and Machine Intelligence*, 24(7): 971–987, 2002.

143. A. Oliva and A. Torralba. Modeling the shape of the scene: A holistic representation of the spatial envelope. *International Journal of Computer Vision*, 42(3):145–175, 2001.

144. I. Otsuka, R. Radhakrishnan, M. Siracusa, and A. Divakaran. An enhanced video summarization system using audio features for a personal video recorder. *IEEE Transactions on Consumer Electronics*, 52(1):168–172, 2006.

145. Y.-F. Pan, X. Hou, and C.-L. Liu. A hybrid approach to detect and localize texts in natural scene images. *IEEE Transactions on Image Processing*, 20(3):800–813, 2011.

146. J. Y. Y. Peng and J. Xiao. Color-based clustering for text detection and extraction in image. In *Proceedings of the 15th International Conference on Multimedia*, pp. 847–850, 2007.

147. X. Peng, S. Setlur, and V. Geovindaraju. Markov random field based binarization for hand-held devices captured document images. In *Proceedings of the Fifth Indian Conference on Computer Vision, Graphics and Image Processing*, pp. 71–76, 2010.

148. X. Peng, S. Setlur, and V. Geovindaraju. An MRF model for binarization of natural scene text. In *Proceedings of the 2011 International Conference on Document Analysis and Recognition*, pp. 11–16, 2011.

149. A. Prest, C. Leistner, J. Civera, C. Schmid, and V. Ferrari. Learning object class detectors from weakly annotated video. In *IEEE Computer Society Conference on Computer Vision and Pattern Recognition*, pp. 3282–3289, 2012.

150. R. Lienhart. Comparison of automatic shot boundary detection algorithm. In *Proceedings of SPIE Storage and Retrieval for Image and Video Databases VII*, pp. 290–301, 1999.

151. Z. Rasheed and M. Shah. Scene detection in Hollywood movies and TV shows. In *Proceedings of IEEE Conference on Computer Vision and Pattern Recognition*, pp. 343–350, 2003.

152. Z. Rasheed and M. Shah. Detection and representation of scenes in videos. *IEEE Transactions on Multimedia*, 7(6), 2005.

153. Z. Rasheed and M. Shah. Scene detection in Hollywood movies and TV shows. In *Proceedings of 2003 IEEE Computer Society Conference on Computer Vision and Pattern Recognition*, vol. 2, pp. 343–348, 2003.

154. Z. Rasheed and M. Shah. Detection and representation of scenes in videos. *IEEE Transactions on Multimedia*, 7(6):1097–1105, 2005.

155. W. Ren, M. Sharma, and S. Singh. Automated video segmentation. In *Proceedings of 2001 International Conference on Information, Communication, and Signal Processing*, pp. 1–11, 2001.

156. Y. Rui, A. Gupta, and A. Acero. Automatically extracting highlights for TV baseball programs. In *Proceedings of the 8th ACM International Conference on Multimedia*, pp. 105–115, 2000.

157. Y. Rui, T. S. Huang, and S. Mehrotra. Constructing table-of-content for videos. *Multimedia Systems*, 7(5):359–368, 1999.

158. T. Mochizuki, R. Ando, and K. Shinoda. A robust scene recognition system for baseball broadcast using data-driven approach. In *Proceedings of the 6th ACM International Conference on Image and Video Retrieval*, pp. 186–193, 2007.

159. S. Di Zenzo. A note on the gradient of a multi-image. *IEEE Transactions on Multimedia*, 9(5):1037–1048, 2007.

160. M. S. Lew. *Princi—plese of Visual Information Retrieval*. Springer Verlag, Berlin, 2001.

161. E. M. Saad, M. I. El-Adawy, M. E. Abu-El-Wafa, and A. A. Wahba. A multifeature speech/music discrimination system. In *Proceedings of the IEEE Canadian Conference on Electrical Computer Enginneering*, pp. 1055–1058, 2002.

162. A. Saffari, C. Leistner, M. Godec, and H. Bischof. Robust multi-view boosting with priors. *Proceedings of European Conference on Computer Vision*, pp. 776–789, 2010.

163. S. Satoh, Y. Nakamura, and T. Kanade. Name-it: Naming and detecting faces in news videos. *IEEE Multimedia*, 6(1):22–35, 1999.

164. F. Schaffalitzky and A. Zisserman. Automated location matching in movies. *Computer Vision and Image Understanding*, 92(2–3):236–264, 2003.

165. B. Scholkopf, R. Herbrich, and A. J. Smola. A generalized representer theorem. In *Proceedings of the Annual Conference on Computational Learning Theory*, pp. 416–426, 2001.

166. S. Lefévre, J. Holler, and N. Vincent. A review of real-time segmentation of uncompressed video sequences for content-based search and retrieval. *Real-Time Imaging*, 9(1):73–98, 2003.

167. J. B. Shi and J. Malik. Normalized cuts and image segmentation. *IEEE Transactions on Pattern Analysis and Machine Intelligence*, 22(8):888–905, 2000.

168. A. Shrivastava, S. Singh, and A. Gupta. Constrained semi-supervised learning using attributes and comparative attributes. In *12th European Conference on Computer Vision*, pp. 369–383, 2012.

169. A. Singh. *Optic Flow Computation: A Unified Perspective*. IEEE Computer Society Press, Washington, D.C., 1992.

170. J. R. Smith and S.-F. Chang. Visually searching the web for content. *IEEE Multimedia Magazine*, 4(3):12–20, 1997.

171. P. Smolensky. *Parallel Distributed Processing*, vol. 1, ch. 6, pp. 198–281. MIT Press, Cambridge, MA, 1986.

172. H. J. Zhang and S. W. Smoliar. Content-based video indexing and retrieval. *Journal of IEEE Multimedia*, 1(2):62–72, 1994.

173. C.-W. Su, H.-Y. Liao, H.-R. Tyan, K.-C. Fan, and L. Chen. A motion-tolerant dissolve detection algorithm. *IEEE Transactions on Multimedia*, 7(6):1106–1113, 2005.

174. Q.-S. Sun, S.-G. Zeng, P.-A. Heng, and D.-S. Xia. The theory of canonical correlation analysis and its application to feature fusion. *Chinese Journal of Computers*, 28(9), 2005.

175. S. Sun. Multi-view Laplacian support vector machines. In *Proceedings of the 7th International Conference on Advanced Data Mining and Applications*, pp. 209–222, 2011.

176. H. Takahashi and M. Nakajima. Region graph based text extraction from outdoor images. In *Proceedings of the Third International Conference on Information Technology and Applications*, pp. 680–685, 2005.

177. Y. P. Tan and H. Lu. Model-based clustering and analysis of video scenes. In *Proceedings of IEEE International Conference on Image Processing*, pp. 617–620, 2002.

178. M. Tapaswi, M. Bäuml, and R. Stiefelhagen. "Knock! Knock! Who is it?" Probabilistic person identification in TV series. In *Proceedings of IEEE Conference on Computer Vision and Pattern Recognition (CVPR)*, pp. 2658–2665, Providence, RI, 2012.

179. R. Tapu and T. Zaharia. A complete framework for temporal video segmentation. In *Proceedings of 2011 IEEE International Conference on Consumer Electronics*, pp. 156–160, Berlin, January 2011.

180. W. Tavanapong and J. Zhou. Shot clustering techniques for story browsing. *IEEE Transactions on Multimedia*, 6(4):517–527, 2004.

181. B. T. Truong, S. Venkatesh, and C. Dorai. Neighborhood coherence and edge based approaches to film scene extraction. In *Proceedings of 2002 International Conference on Pattern Recognition*, pp. 350–353, 2002.

182. B. T. Truong and S. Venkatesh. Video abstraction: A systematic review and classification. *ACM Transactions on Multimedia Computing, Communications and Applications*, 3(1):1–37, 2007.

183. H. D. Wactlar, E. G. Hauptmann, and M. J. Witbrock. Informedia: News-on-demand experiments in speech recognition. In *Proceedings of ARPA Speech Recognition Workshop*, pp. 18–21, 1996.

184. J. Wang. Broadcast and television information cataloguing specification: Part 1: Specification for the structure of teleplays. Technical Report 202.1. China's State Administration for Radio, Film and Television, 2004.

185. J. Wang, L. Duan, H. Lu, J. S. Jin, and C. Xu. A mid-level scene change representation via audiovisual alignment. In *Proceedings of 2006 IEEE International Conference on Acoustics, Speech and Signal Processing*, vol. 2, pp. 409–412, 2006.

186. R. Wang, W. Jin, and L. Wu. A novel video caption detection approach using multi-frame integration. In *Proceedings of the 17th International Conference on Pattern Recognition*, pp. 449–452, 2004.

187. S. Wang, H. Lu, F. Yang, and M. H. Yang. Superpixel tracking. In *Proceedings of the 13th International Conference on Computer Vision*, pp. 1323–1330, 2011.

188. D. Wijesekera and D. Barbara. Mining cinematic knowledge work in progress. In *Proceedings of International Workshop on Multimedia Data Mining (MDM/KDD)*, pp. 98–103, 2000.

189. J. Wu and J. M. Rehg. Where Am I? Place instance and category recognition using spatial pact. In *Proceedings of 2008 IEEE Computer Society Conference on Computer Vision and Pattern Recognition*, pp. 1–8, 2008.

190. J. Wu and J. M. Rehg. Centrist: A visual descriptor for scene categorization. *IEEE Transactions on Pattern Analysis and Machine Intelligence*, 33(8):1489–1501, 2011.

191. D. Xia, X. Deng, and Q. Zeng. Shot boundary detection based on difference sequences of mutual information. In *Proceedings of the Fourth International Conference on Image and Graphics*, pp. 389–394, Chengdu, China, August 2007.

192. J. Xiao, J. Hays, K. A. Ehinger, A. Oliva, and A. Torralba. Sun database: Large-scale scene recognition from abbey to zoo. In *Proceedings of 2010 IEEE Conference on Computer Vision and Pattern Recognition*, pp. 3485–3492, 2010.

193. J. X. Xiao, K. A. Ehinger, A. Oliva, and A. Torralba. Recognizing scene viewpoint using panoramic place representation. In *Proceedings of the 2012 IEEE Conference on Computer Vision and Pattern Recognition*, pp. 2695–2702, 2012.

194. L. Xie, S. F. Chang, H. Sun, and A. Divakaran. Structure analysis of soccer video with hidden Markov models. In *Proceedings of the International Conference on Acoustic, Speech, and Signal Processing*, pp. 4096–4099, 2002.

195. C. Xiong, G. Y. Gao, Z. J. Zha, S. C. Yan, H. D. Ma, and T. K. Kim. Adaptive learning for celebrity identification with video context. *IEEE Transactions on Multimedia*, 16(5):1473–1485, 2013.

196. W. Xiong and J. C.-M. Lee. Efficient scene change detection and camera motion annotation for video classification. *Computer Vision and Image Understanding*, 71(2):166–181, 1998.

197. Z. Xiong, R. Radhakrishnan, A. Divakaran, and T. S. Huang. Highlights extraction from sports video based on an audio-visual marker detection framework. In *Proceedings of the IEEE International Conference on Multimedia and Expo*, pp. 1613–1616, 2005.

198. M. Xu, L. Duan, L. Chia, and C. Xu. Audio keyword generation for sports video analysis. In *Proceedings of the 12th Annual ACM International Conference on Multimedia*, pp. 1055–1058, 2002.

199. Q. Xu, Y. Liu, X. Li, Z. Yang, J. Wang, M. Sbert, and R. Scopigno. Browsing and exploration of video sequences: A new scheme for key frame extraction and 3D visualization using entropy based Jensen divergence. *Information Sciences*, 278(10):736–756, 2014.

200. R. Yan, J. Zhang, J. Yang, and A. G. Hauptmann. A discriminative learning framework with pairwise constraints for video object classification. *IEEE Transcations on Pattern Analysis and Machine Intelligence*, 28, 2006.

201. Y. Yang, S. Lin, Y. Zhang, and S. Tang. Highlights extraction in soccer videos based on goal-mouth detection. In *Proceedings of the 9th International Symposium on Signal Processing and Its Applications*, pp. 1–4, 2007.

202. Y. Yang, G. Shu, and M. Shah. Semi-supervised learning of feature hierarchies for object detection in a video. In *IEEE Computer Society Conference on Computer Vision and Pattern Recognition*, pp. 1650–1657, 2013.

203. D. Yarowsky. Unsupervised word sense disambiguation rivaling supervised methods. In *33rd Annual Meeting of the Association for Computational Linguistics*, pp. 189–196, 1995.

204. N. Ye, J. Li, and Z. Zhang. Fast gradual effect scene change detection algorithm in MPEG domain. *Journal of Shanghai Jiaotong University*, 35(1):34–36, 2001.

205. Q. Ye and Q. Huang. A new text detection algorithm in images/video frames. In *Proceedings of 5th Pacific Rim Conference on Multimedia*, pp. 858–865, 2004.

206. B.-L. Yeo. Efficient processing of compressed images and video. PhD thesis, Princeton University, 1996.

207. B.-L. Yeo and B. Liu. Rapid scene analysis on compressed video. *IEEE Transactions on Circuits and Systems for Video Technology*, 5(6):533–544, 1995.
208. M. Yeung, B. Yeo, and B. Liu. Segmentation of videos by clustering and graph analysis. *Computer Vision and Image Understanding*, 71:94–109, 1998.
209. M. M. Yeung and B. Liu. Efficient matching and clustering of video shots. In *Proceedings of International Conference on Image Processing*, pp. 338–341, 1995.
210. D. Zhou and Y. Zhu. Scene change detection based on audio and video content analysis. In *Proceedings of 5th International Conference on Computational Intelligence and Multimedia Applications*, pp. 229–234, 2003.
211. H.-W. Yoo, H.-J. Ryoo, and D.-S. Jang. Gradual shot boundary detection using localized edge blocks. *Multimedia Tools and Applications*, 28(3):283–300, 2006.
212. YouTube. Statistic report of YouTube: https://www.youtube.com/yt/press/statistics.html.
213. J. Yuan, J. Li, F. Lin, and B. Zhang. A unified shot boundary detection framework based on graph partition model. In *Proceedings of the 13th Annual ACM International Conference on Multimedia*, pp. 539–542, Hilton, Singapore, March 2005.
214. X. L. Zeng, W. M. Hu, and W. Liy. Key-frame extraction using dominant-set clustering. In *Proceedings of the 2008 IEEE International Conference on Multimedia and Expo*, pp. 1285–1288, 2008.
215. H. J. Zhang, A. Kankanhalli, and S. W. Smoliar. Automatic partitioning of full-motion video. *Multimedia Systems*, 1(1):10–28, 1993.
216. S. Zhang, Q. Tian, Q. Huang, W. Gao, and S. Li. Utilizing effective analysis for efficient movie browsing. In *Proceedings of the 16th IEEE International Conference on Image Processing*, pp. 1853–1856, 2009.
217. X. Zhang, L. Zhang, X.-J. Wang, and H.-Y. Shum. Finding celebrities in billions of web images. *IEEE Transactions on Multimedia*, 14(4):995–1007, 2012.
218. Y. F. Zhang, C. Xu, H. Lu, and Y. M. Huang. Character identification in feature-length films using global face-name matching. *IEEE Transactions on Multimedia*, 11(9):1276–1288, 2009.
219. Y. Zhang. *Content-Based Visual Information Retrieval*. Science Press, Beijing, China, 2003.
220. M. Zhao, J. Yagnik, H. Adam, and D. Bau. Large scale learning and recognition of faces in web videos. In *8th IEEE International Conference on Automatic Face and Gesture Recognition*, pp. 1–7, 2008.
221. W.-L. Zhao, K. C.-W. Ngo, H.-K. Tan, and X. Wu. Near duplicate keyframe identification with interest point matching and pattern learning. *IEEE Transactions on Multimedia*, 9(5):1037–1048, 2007.
222. Y. Zhao, T. Wang, P. Wang, W. Hu, Y. Du, Y. Zhang, and G. Xu. Scene segmentation and categorization using ncuts. In *Proceedings of 2007 IEEE Conference on Computer Vision and Pattern Recognition*, pp. 1–7, 2007.
223. Z. Zhao, S. Jiang, Q. Huang, and G. Zhu. Highlight summarization in sports video based on replay detection. In *Proceedings of the IEEE International Conference on Multimedia and Expo*, pp. 1613–1616, 2006.
224. Y. Zhong, H. Zhang, and A. Jain. Automatic caption localization in compressed video. *IEEE Transcations on Pattern Analysis and Machine Intelligence*, 22(4):385–392, 2000.
225. D. Zhou, O. Bousquet, T. N. Lal, J. Weston, and B. Scholkopf. In *Proceedings of 17th Neural Information Processing Systems (NIPS)*, pp. 321–328, 2003.
226. X. Zhou, X. Zhuang, H. Tang, M. Hasegawa-Johnson, and T. S. Huang. A novel Gaussianized vector representation for natural scene categorization. In *Proceedings of the 19th International Conference on Pattern Recognition*, pp. 1–4, 2008.

227. S. Zhu and Y. Liu. Automatic scene detection for advanced story retrieval. *Expert Systems with Applications*, 36(3, Part 2):5976–5986, 2009.

228. X. Zhu, J. Lafferty, and Z. Ghahramani. Semi-supervised learning: From Gaussian fields to Gaussian processes. In *International Conference on Machine Learning*, pp. 912–919, 2003.

229. T. Zhuang, Y. Pan, and F. Wu. *Online Multimedia Information Analysis and Retrieval.* Tsinghua University Press, Beijing, China, 2002.

230. C. Zuzana, P. Ioannis, and C. Nikou. Information theory-based shot cut or fade detection and video summarization. *IEEE Transactions on Circuits and Systems for Video Technology*, 16(1):82–91, 2006.

231. I. Laptev, M. Marszalk, C. Schmid, and B. Rozenfeld. Learning realistic human actions from movies. In *Proceedings of IEEE Conference on Computer Vision and Pattern Recognition*, 2008, pp. 1–8.

232. J. Yang, J. Y. Yang, D. Zhang and J. F. Lu. Feature fusion: parallel strategy vs. serial strategy. *Pattern Recognition*, 36(6):1369–1381, 2003.

Index

A

Adaptive resonance theory, 94, 96–97
ADVENT, Columbia University, 4
ART. *See* Adaptive resonance theory
Audiovisual feature extraction,
53–54
Audiovisual information-based highlight
extraction, 119–127
framework, 120–121
hidden Markov model-based approach,
120
unrelated scene removal, 121–126
background differences, 123
scene types, 121–122
score bar change detection, 125
special audio words detection, 126
speech/nonspeech discrimination,
122–123
subshot segmenting, 125
Automatic movie, teleplay cataloguing, 12,
129–138

B

Block matching and motion vector-based
SBD methods, 28
The Broadcast Sportswear Video Retrieval
System, 5

C

CAMSHIFT algorithm. *See* Continuously
adaptive mean shift algorithm
Canonical correlation analysis, 11, 50,
55–56, 68
Carnegie Mellon University, Informedia
project, 4

Categories of key frame extraction, 43–46
clustering-based algorithms, 44
optimization-based algorithms, 46
sequential algorithms, 43–44
shot-based algorithms, 45–46
CBVR. *See* Content-based video retrieval
CCA. *See* Canonical correlation analysis
Census transform histogram, video scene
recognition, 77
Central China Normal University, 5
CENTRIST. *See* Census transform
histogram
Character identification, 3, 9–12, 91–118
adaptive learning, 96–105
classification error bound, related
LapSVM, 104–105
initial learning, 97
Laplacian support vector machine,
100–101
learning curves, 115–116
related LapSVM, 101–104
adaptive resonance theory, 94, 96
clusters, 109
database construction, 106–110
duration, 109
experimental settings, 110–111
experiments, 106–118
face recognition, 110
Gabor feature, 110
image data, 106
Kernel Hilbert Space, reproducing, 96,
101
local binary pattern feature, 110
optical character recognition, 10
scale-invariant feature transform feature,
110
semisupervised learning, 113–115

supervised learning, 111–113
transductive support vector machine, 113
video constraint in graph, 111
video data, 106–110
YouTube celebrity dataset, 116–118
China Academy of Science, 5
Classification error bound related LapSVM, 104–105
Clustering-based algorithms, 44
Clusters, 7, 71, 77, 109
character identification, 109
Color histogram-based shot boundary detection methods, 28
Columbia University, ADVENT, 4
Computer Vision and Pattern Recognition, 19
Connected component, 9, 62, 68
Content-based video retrieval, 4–5, 131
Continuously adaptive mean shift algorithm, video text detection, 62
Corner distribution-based MCFB removal, 34–37
CVPR. *See* Computer Vision and Pattern Recognition

D

DCT Discrete Cosine Transform, 6
Detection, video text, 62–70
adjacent frames, motion block attribute on, 64–68
continuously adaptive mean shift algorithm, 62
motion block attribute of consecutive frames, clustering block based on, 66–67
motion perception field, 63–64
constructing on consecutive frames, 67–68
multiframe verification, 69
support vector machine classifier, 62
Direct scale-invariant feature transform feature extraction-based scene recognition, 87–88
Dirichlet analysis, 76
Discrete cosine transform coefficient, direct current component, 6
Duminda, George Mason University, 5

E

Edge-based shot boundary detection methods, 27–28
Extraction, video text, 70–73
with k means, 71–73

F

Face recognition, character identification, 110
Feature learning, 22–23
Gaussian mixture models, 23
sparse autoencoder, 22–23
sparse restricted Boltzmann machine, 23
Focus region, 11, 27, 29–30, 37, 39, 52, 109
Frame difference calculation, 29–33
Frame difference measurement, 30–33
Frame difference scene detection algorithm, 7
Frame similarity measure with mutual information, 31–33
French National Institute for Research in Computer Science and Control, 19

G

Gabor feature, 17–19, 23, 110
Gaussian mixture models, 23
General design of demo, 132–138
George Mason University, Duminda, 5
GMMs. *See* Gaussian mixture models
Gradual transitions, 6, 25, 31, 52
Graph-based shot similarity methods, 7
Gray value-based shot boundary detection methods, 27
GTs. *See* Gradual transitions

H

Hidden Markov model, 6, 77, 120, 126
Highlight extraction, audiovisual information-based, 119–127
framework, 120–121
hidden Markov model-based approach, 120
unrelated scene removal, 121–126
background differences, 123

scene types, 121–122
score bar change detection, 125
special audio words detection, 126
speech/nonspeech discrimination, 122–123
subshot segmenting, 125
Histogram of oriented gradients, 19–20, 23
HMM. *See* Hidden Markov model
HOG. *See* Histogram of oriented gradients

I

Informedia project, Carnegie Mellon University, 4
Initial learning, 94, 97, 100
INRIA. *See* French National Institute for Research in Computer Science and Control

J

Joint entropy, 32

K

KCCA. *See* Kernel canonical correlation analysis
Kernel canonical correlation analysis, 50–60, 133
  feature fusion-based method, 52–58
    audiovisual feature extraction, 53–54
    graph cut, scene detection based on, 56–58
    key frame extraction, 52–53
    spectrum flux, 54
    zero-crossing rate, 54
  merging-based scene detection, 51
  model-based scene detection, 51
  splitting-based scene detection, 51
Kernel Hilbert Space reproduction, character identification, 96, 101
Key frame extraction, 41–48, 52–53
  categories, 43–46
    clustering-based algorithms, 44
    optimization-based algorithms, 46
    sequential algorithms, 43–44
    shot-based algorithms, 45–46
  fixed as known prior, 42–43

kernel canonical correlation analysis, 52–53
key frame set size, 42–43
left as unknown posterior, 43
using panoramic frame, 46–48
Key point descriptor, 14, 17
Key point localization, 14–16
Key Technology and Application of Media Service Collaborative Processing, 130

L

Laplacian support vector machine, 100–101
  character identification, 100–101
Latent Dirichlet analysis, 76, 78, 81–83
  video scene recognition, 78, 81
    scene classification using, 81–83
LBP. *See* Local binary pattern
LDA. *See* Latent Dirichlet analysis
Learning curves, character identification, 115–116
Local binary pattern, 21–23, 110
  character identification, 110

M

Main function modules, 133–136
Main function of demo, 131
Markov chain Monte Carlo, video scene recognition, 83
Maximally stable extremal regions, 20–21, 23
  extremal region, 21
  image, 20
  outer region boundary, 21
  region, 21
MCMC. *See* Markov chain Monte Carlo
Media asset management area, 130
MediaMill, 4
Merging-based scene detection, kernel canonical correlation analysis, 51
Metadata entry module, 134–135
MFV. *See* Multiframe verification
Microsoft, 5, 133
Model-based scene detection, 51
  kernel canonical correlation analysis, 51

Motion perception field, 9, 63–64, 66–69
Movie, teleplay characters recognition
module, 135–136
Movie, teleplay production and
broadcasting system, 130
MPF. *See* Motion perception field
MSERs. *See* Maximally stable extremal
regions
Multiframe verification, video text
detection, 69
Multimodality movie scene detection,
49–60
experiment, 58–59
kernel canonical correlation analysis,
50–59
feature fusion-based method, 52–58
graph cut, scene detection based on,
56–58
merging-based scene detection, 51
model-based scene detection, 51
splitting-based scene detection, 51
Mutual information, 7, 11, 27, 29–33, 37,
53, 77
Mutual information maximization, video
scene recognition, 77

N

NewBR. *See* The News Video Browsing
and Retrieval System
The News Video Browsing and Retrieval
System, 5
News Video Browsing System, 5
NVBS. *See* News Video Browsing System

O

OCR. *See* Optical character recognition
Optical character recognition, 10, 61–62,
70, 135
Optimization-based algorithms, 46
Orientation assignment, 14, 16–17
Oriented gradients, histogram, 19–20
Computer Vision and Pattern
Recognition, 19
French National Institute for Research
in Computer Science and Control,
19
Overall diagram of demo, 132–133

P

Panoramic frame
key frame extraction using, 46–48
video scene recognition, performance
evaluation, 86–87
Patch extraction, representative feature,
video scene recognition, 79–81
PNN. *See* Probabilistic neural network
Probabilistic neural network, video scene
recognition, 77
Proposed systems, 4

Q

QBIC. *See* Query by Image Content
Query by Image Content project, 4

R

Removal of scenes, 121–126
background differences, 123
scene types, 121–122
score bar change detection, 125
special audio words detection, 126
speech/nonspeech discrimination,
122–123
subshot segmenting, 125
Reproducing Kernel Hilbert Space,
96, 101
Research, 3–12
ADVENT, 4
The Broadcast Sportswear Video
Retrieval System, 5
Central China Normal University, 5
CENTRIST feature descriptor, 8
character identification, 9–12
optical character recognition, 10
China Academy of Science, 5
content-based video retrieval, 4
CueVideo, 4
discrete cosine transform coefficient,
direct current component, 6
Duminda, George Mason University, 5
hidden Markov model, 6
Informedia, 4
MediaMill, 4
Microsoft, 5
The News Video Browsing and Retrieval
System, 5

News Video Browsing System, 5
proposed systems, 4
QBIC, 4
Query by Image Content project, 4
scene detection, recognition, 7–9
shot boundary detection, 6–7, 11
Taiwan National Chiao Tung
    University, 5
traditional video scene detection
    methods:, 7
    frame difference scene detection
        algorithm, 7
    graph-based shot similarity
        methods, 7
    visual features clustering algorithm, 7
video scene segmentation, recognition,
    11–12
video text recognition, 9, 12
    connected component, 9
    motion perception field, 9
RKHS. *See* Reproducing Kernel Hilbert
    Space
Running environment, 133

S

SBD. *See* Shot boundary detection
Scale-invariant feature transform feature,
    character identification, 110
Scale-space extrema, 14–15
Scene detection, recognition, 7–9
Scene detection module, 135
Scene recognition, 75–89
    census transform histogram, 77
    direct scale invariant feature transform
        feature extraction-based method,
        87
    enhanced recognition based on,
        83–85
        experimental conditions, 87–89
        methods compared, 85–86
        panoramic frames, performance
            evaluation, 86–87
    latent Dirichlet analysis, 78, 81
        scene classification using, 81–83
    Markov chain Monte Carlo, 83
    mutual information maximization, 77
    patch extraction, representative feature,
        79–81

probabilistic neural network, 77
    video segmentation, 79
Scene recognition module, 135
Scene removal, 121–126
    background differences, 123
    scene types, 121–122
    score bar change detection, 125
    special audio words detection, 126
    speech/nonspeech discrimination,
        122–123
    subshot segmenting, 125
Segmentation, video scene recognition,
    79
Semisupervised learning, character
    identification, 113–115
Sequential algorithms, 43–44
Shannon entropy, 31
Short-time energy, 11, 52, 54
Shot-based algorithms, 45
Shot boundary detection, 6–7, 11, 25–40
    color histogram-based SBD
        methods, 28
    edge-based SBD methods, 27–28
    gray value-based SBD methods, 27
Shot boundary detection module, 135
Shot similarity graph, 7, 11, 51–52,
    57, 60
SIFT. *See* Scale-invariant feature transform
Size of key frame set, 42–43
Sparse autoencoder, 22–23
Sparse restricted Boltzmann machine, 23
Spectrum flux, 54
    kernel canonical correlation analysis, 54
Splitting-based scene detection, 51
    kernel canonical correlation analysis, 51
SPORTBB, 5
SSG. *See* Shot similarity graph
STE. *See* Short-time energy
SUM. *See* Support vector machine classifier
Superimposed text detection, video text
    detection, 68–70
Supervised learning, character
    identification, 111–113
Support vector machine classifier, 3, 17,
    20, 44, 62, 65
SVM. *See* Support vector machine
SVM classifier. *See* Support vector machine
    classifier
System design and realization, 136–138

T

Taiwan National Chiao Tung University, 5
Temporal redundant frame reduction,
    33–34
Traditional video scene detection methods,
    7
    frame difference scene detection
        algorithm, 7
    graph-based shot similarity methods, 7
    visual features clustering algorithm, 7
Transductive support vector machine, 113,
    115
    character identification, 113
TSVM. *See* Transductive support vector
    machine

U

Unknown posterior, 42–43
Unrelated scene removal, 121–126
    background differences, 123
    scene types, 121–122
    score bar change detection, 125
    special audio words detection, 126
    speech/nonspeech discrimination,
        122–123
    subshot segmenting, 125

V

Video abstraction, 136
Video cataloguing, 137–138
Video indexing and retrieval, 136–137
Video scene recognition, 75–89
    census transform histogram, 77
    direct scale invariant feature transform
        feature extraction-based method,
        87
    enhanced recognition based on, 83–85
        experimental conditions, 87–89
        methods compared, 85–86
        panoramic frames, performance
            evaluation, 86–87
    latent Dirichlet analysis, 78, 81
        scene classification using, 81–83
    Markov chain Monte Carlo, 83
    mutual information maximization, 77

patch extraction, representative feature,
    79–81
probabilistic neural network, 77
video segmentation, 79
Video scene segmentation, recognition,
    11–12
Video text detection, recognition, 61–73
    video text detection, 62–70
        adjacent frames, motion block
            attribute on, 64–68
        constructing on consecutive frames,
            67–68
        continuously adaptive mean shift
            algorithm, 62
        motion block attribute of consecutive
            frames, 66–67
        motion perception field, 63–64
        multiframe verification, 69
        superimposed text detection, 68–70
        support vector machine classifier, 62
    video text extraction, 70–73
        with *k* means, 71–73
Video text recognition, 2, 9, 12, 61–62,
    135
    connected component, 9
    motion perception field, 9
Video text recognition module, 135
Visual features clustering algorithm, 7
Visual features extraction, 13–23
    feature learning, 22–23
        Gaussian mixtures, 23
        sparse autoencoder, 22–23
        sparse restricted Boltzmann machine,
            23
    Gabor feature, 17–19
    local binary pattern, 21–22
    maximally stable extremal regions,
        20–21
        extremal region, 21
        image, 20
        outer region boundary, 21
        region, 21
    oriented gradients, histogram, 19–20
        Computer Vision and Pattern
            Recognition, 19
        French National Institute for
            Research in Computer Science and
            Control, 19

scale-invariant feature transform, 14–17
    key point descriptor, 17
    key point localization, 15–16
    orientation assignment, 16–17
    scale-space extrema, 14–15

W

Words, audio, detection of, 126

Y

YouTube celebrity dataset, 116–118

Z

ZCR. *See* Zero-crossing rate
Zero-crossing rate, kernel canonical
    correlation analysis, 54